Computerized Data Acquisition and Analysis for the Life Sciences

This book offers an excellent introduction to computer-based data acquisition and analysis and will be especially valuable to those with little or no engineering background. In an informal, easy-to-read style, it guides researchers through the basics of data acquisition systems, explains the important concepts underlying sampled data, and gives numerous examples of how to analyse the recorded information. While aimed at researchers in the life sciences, the topics covered are general and will be valuable to anyone interested in learning to use data acquisition systems.

A variety of high-quality hardware and software is now available to researchers, and computerized data acquisition systems are often the principal method of recording experimental results. Yet for the non-engineer, setting up these systems can be a difficult task. This book takes the reader through the process step by step, from the type of input to be used (single-ended or differential, unipolar or bipolar) through to the sampling rate and the size of the resultant data files. It explains how to set up preamplifiers to get the best results, and covers the main types of transducer encountered in life-science work. It then discusses how to obtain useful information from the large amounts of data recorded. The principles can be applied to the collection of data from respiratory apparatus, intracellular and extracellular electrodes, electrocardiograms, behavioural-science experiments, kinematics, and a host of other situations. Many illustrations and worked examples accompany the text, unfamiliar terms are explained, and the mathematics are kept as simple as possible.

This book is an invaluable tool for the non-engineer who is collecting and analysing experimental data using data acquisition systems. Researchers, graduate students, and technicians will find it an up-to-date, accessible, and indispensable guide for setting up their equipment and getting the most out of their experimental data.

Simon S. Young is a Senior Principal Scientist at the Schering-Plough Research Institute in Kenilworth, New Jersey.

Computerized Data Acquisition and Analysis for the Life Sciences

A Hands-on Guide

SIMON S. YOUNG
Schering-Plough Research Institute

CAMBRIDGE
UNIVERSITY PRESS

PUBLISHED BY THE PRESS SYNDICATE OF THE UNIVERSITY OF CAMBRIDGE
The Pitt Building, Trumpington Street, Cambridge, United Kingdom

CAMBRIDGE UNIVERSITY PRESS
The Edinburgh Building, Cambridge CB2 2RU, UK
40 West 20th Street, New York, NY 10011-4211, USA
10 Stamford Road, Oakleigh, VIC 3166, Australia
Ruiz de Alarcón 13, 28014 Madrid, Spain
Dock House, The Waterfront, Cape Town 8001, South Africa

http://www.cambridge.org

First published 2001

Printed in the United States of America

Typeface Melior 10/13.5 pt   *System* 3B2   [KW]

*A catalogue record for this book is available from
the British Library*

Library of Congress Cataloging-in-Publication Data
Young, Simon S., 1958–
   Computerized data acquisition and analysis for the life sciences: a hands-on guide/
Simon S. Young
      p.   cm.
   Includes bibliographical references
   ISBN 0-521-56281-3 (hardback)
   ISBN 0-521-56570-7 (paperback)
   1. Life sciences – Data processing.   2. Automatic data collection systems.   I. Title.
   QH324.2.Y68   2001
   570′.285 – dc21   00-045436

ISBN 0 521 56281 3 hardback
ISBN 0 521 56570 7 paperback

Contents

Preface

There has been a small revolution in data acquisition systems for scientific use in recent years. It was not so long ago that hardware and processing power were expensive and software nonexistent. Every program had to be written from scratch in a low-level language and, if you did not want your program to take all night to run, it usually included large chunks of assembler code as well. Nowadays there are many data acquisition systems to choose from, with good hardware and vastly improved software running on fast personal computers. Many such systems are purchased with the aim of using them as the primary method of recording experimental data in the laboratory. Unfortunately, it is not always as easy to set up the system as the manufacturers would have us believe and getting the machine to do what you want it to do with your apparatus can be a daunting and frustrating task. This book was written to help people collect and analyse experimental data using digital data acquisition systems and is particularly for those whose field of expertise is not engineering or computing. The book explains how data acquisition systems work, how to set them up to obtain the best performance and how to process the data. Terms which may be unfamiliar are explained so that manufacturer's specifications and literature become clearer and the differences between systems can be appreciated. The topics covered in the book are general but the examples are slanted towards the life sciences because this is often the discipline in which people have the greatest trouble setting up data acquisition systems. For people with an engineering background some of the material will be familiar but there are still useful tips and short cuts.

Program listings are not included in the book. There are some practical problems associated with publishing programs: they have to be cross-platform and written in several languages if they are to be useful to everyone. But the main reason is that I would strongly urge you to forget about low-level coding altogether. Buy a high-level language designed for signal processing (LabView,

IDL, Matlab, etc.) and spend your programming time assembling algorithms from the building blocks that these programs provide, not reinventing the wheel. If you absolutely have to write code, use a library of prewritten functions (such as those sold by National Instruments) that have been tested and debugged for you.

This book would not have been possible without the help and encouragement of many people. The original idea arose out of the author's work with graduate students at the Universities of Cambridge, Guelph and Waterloo when it became clear that there was a need for a book on data acquisition and analysis written for non-engineers. Dr. Robin Gleed provided the initial encouragement that launched the project. Data for some of the examples used in Chapter 6 were provided by Dr. Jeff Thomason, Dr. Janet Douglas, and Dr. Zoe Dong. Dr. Richard Hughson ran the research project that led to the development of TSAS. Dr. Craig Johnson reviewed early versions of the manuscript and provided data for some of the examples. Dr. Robin Smith and Dr. Michael Penn of Cambridge University Press were invaluable at all stages of writing and editing the book. And most of all, I am indebted to my wife, Janet Douglas, for endless patience, encouragement, and manuscript reviews.

chapter one

The Bare Essentials

This chapter outlines the key tasks associated with setting up and using a data acquisition system. It is essentially a checklist with pointers to the sections of the book that cover the topics in more detail.

DEFINE THE VARIABLES TO BE MEASURED

The first task is to decide what specific physiological effects are to be recorded. For example, suppose the effects of a drug on cardiovascular function are being investigated. There is quite a range of variables that can be used as indicators of cardiovascular function: heart rate, blood pressure, cardiac output, systemic vascular resistance, rate of change of pressure (dp/dt), end diastolic volume, ejection fraction, myocardial fibre shortening, etc. Some of these are relatively easy to measure and some are quite involved. Often the starting point is to replicate a published method and this considerably shortens the development process.

CONVERT THE PHYSIOLOGICAL EFFECT TO A VOLTAGE

All data acquisition systems fundamentally only measure one thing: voltage. A transducer is needed (Chapter 4) to convert the variable(s) of interest (e.g. vascular pressure) into a representative voltage. Each transducer will also need an amplifier, and this is usually specific for the particular type of transducer. Even if biopotentials (which are already voltages) are being measured, one cannot just hook up an electrode to the input of a data acquisition system – it is very unlikely to work well. An amplifier specific for the type of biopotential (intracellular potential, ECG, EEG, EMG, etc.) is needed in order to obtain usable results. In addition, if recordings are being made from human subjects, an isolation amplifier is essential for safety.

SCALE THE SIGNAL

The main analogue-to-digital converter (ADC) has a range of voltages that it can accept. This range is typically ± 10 V, ± 5 V, or ± 1 V and nearly all systems allow the range to be changed. The aim is to use a good part of this range (giving good resolution) without exceeding it (which severely distorts the signal). Matching the signal range to the ADC range is performed by setting the gain of the transducer amplifier so that the maximum expected signal is about half or two thirds of the ADC range. For example, if an ECG signal is being recorded, the maximum amplitude expected is about 3 mV (the height of a large R wave). An ECG preamplifier typically has a gain of $\times 1000$, giving a maximum output of 3 V. Setting the ADC range to ± 5 V would be a good choice, although even ± 10 V would not be bad. Another example is a blood pressure transducer. These typically have a range of 0–300 mmHg, with a corresponding output from their amplifier of 0–10 V. Setting the ADC range to ± 10 V will use the full range of the transducer. If the blood pressure exceeds 300 mmHg, which is unlikely, the transducer will not give a reliable reading, so using the full range of the ADC is fine.

SELECT A UNIPOLAR OR BIPOLAR INPUT RANGE

For most signals the input should be set to bipolar. If the input is set to bipolar for all signals, the system will function normally. The only disadvantage is that there is a potential loss of resolution if the signal is guaranteed never to be negative. In such cases setting the input to unipolar mode will improve the resolution by a factor of up to 2. On the other hand, if the input is set to unipolar and the signal goes negative, it will be severely distorted by the recording process. Chapter 2 has the details.

CHOOSE SINGLE-ENDED OR DIFFERENTIAL INPUTS

In most cases the inputs should be set to single-ended. Differential inputs are better at rejecting noise, particularly if the signal is small or if it is connected to the data acquisition system by long cables. On the other hand, differential inputs are more complex to wire up and in differential mode you usually only have half the number of inputs available compared to single-ended mode. Again, Chapter 2 has the details.

TABLE 1.1 **SUGGESTED SAMPLING RATES FOR DIFFERENT TYPES OF BIOLOGICAL SIGNALS**

Signal being recorded	Sampling rate (Hz)
Temperature, ionic composition, other variables that do not change much over a period of a few seconds	5–20
Blood pressure (except intravascular catheter-tip transducers), gas composition (using a fast analyser such as mass spectrometer), respiratory gas temperature, joint angle (goniometry), ECG, EEG	200–400
EMG, evoked potentials, action potentials	600–1,000
Intravascular catheter-tip transducers; all studies involving impacts such as force plate studies	2,000–3,000
Action potentials, transmembrane currents (patch and intracellular)	10,000–20,000

CHOOSE THE SAMPLING RATE AND ANTI-ALIASING FILTERS

Setting the sampling rate too low will distort the data ('aliasing'). Setting it too high will not do any harm, the data will just take up more storage space. Chapter 3 goes into the details. For now, Table 1.1 gives some very rough guidelines for sampling rates for different types of biological signals.

If possible, try to include a low-pass filter (an 'anti-aliasing' filter) in between the transducer amplifier and the ADC. The 'corner frequency' of the filter should be no higher than about 1/4 of the sampling rate: if the sampling rate is 400 samples/s, for example, set the filter corner frequency to below 100 Hz.

SIZE OF DATA FILES

The amount of data that will be collected can be calculated from the following equation:[1]

$$\text{amount of data (bytes)} = 2 \times \text{no. of channels} \times \text{sampling rate} \times \text{recording length (seconds)}$$

For example, recording from four channels at 200 Hz for 30 min will generate about 2.7 Mbyte of data. Some means of storing and archiving these data is needed.

CALIBRATION

Nearly all transducers require calibration before use. The calibration process relates the voltage recorded by the data acquisition system to the quantity measured by the transducer (pressure, temperature, etc.). The vast majority of transducers and transducer/amplifiers are *linear*. This means that a plot of the voltage recorded by the data acquisition system against the quantity being measured is a straight line (Figure 1.1). A linear system is particularly easy to calibrate because the straight line relating voltage to physical effect can be defined by just two points, and such a calibration process is known as a *two-point calibration*. Nonlinear transducers require a more complex calibration procedure, which is covered in Chapter 4.

To perform a two-point calibration, two known values of the variable (spanning most of the range of interest) are applied to the transducer with a reference system. The corresponding voltage is recorded both times. For example, a temperature transducer that is to be used to record mammalian body temperature might be calibrated at 35°C and 45°C. Two water baths at approximately these temperatures are set up. The exact temperature of each bath is measured with a precision thermometer and at the same time the voltage produced by the transducer (the latter is usually taped to the thermometer for this process) is recorded. Two corresponding pairs of values (temperature and voltage) are thus recorded: T_{hi}, T_{low} and V_{hi}, V_{low}. A calibration graph (Figure 1.2) now has to be constructed from these two points so that any voltage can be converted back into the corresponding temperature.

The straight line relating temperature and voltage (Figure 1.2) is usually defined by its *scale* and *offset*. The general equation of a line is

$$y = mx + c$$

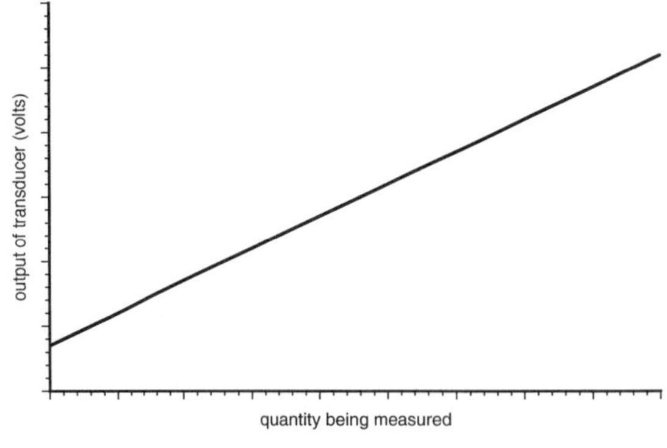

Figure 1.1 The response of a linear transducer and amplifier. A plot of the output of the transducer (in V) against the quantity being measured (temperature, pressure, etc.) is a straight line if the transducer is linear.

Figure 1.2 Two-point calibration of a linear temperature transducer. Two temperature points are selected that span the range of interest, and the output of the transducer at each temperature is measured. The scale and offset of the line joining the two points are calculated, and from this any measured voltage can be converted into a corresponding temperature. Note that the graph has temperature (the independent variable) on the y-axis rather than the more usual x-axis.

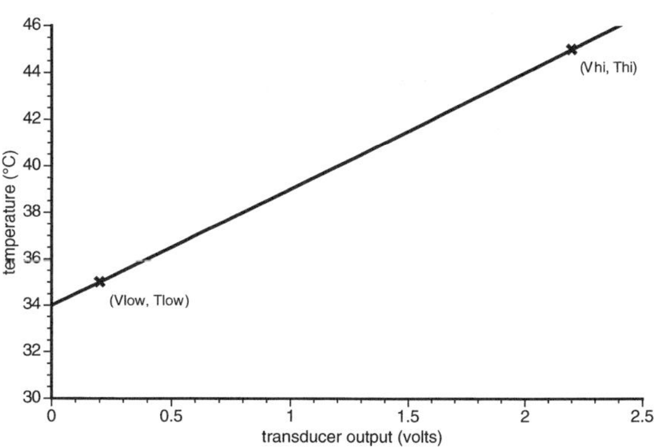

where m is the scale and c the offset. For the calibration graph (Figure 1.2) the equation of the line is given by the following equation:[2]

$$\text{temperature} = (\text{scale} \times \text{voltage}) + \text{offset}$$

The scale and offset are calculated from

$$\text{scale} = \frac{T_{\text{hi}} - T_{\text{low}}}{V_{\text{hi}} - V_{\text{low}}}$$

$$\text{offset} = T_{\text{hi}} - (\text{scale} \times V_{\text{hi}})$$

Now any voltage V can be converted to a corresponding temperature T with

$$T = (\text{scale} \times V) + \text{offset}$$

A data acquisition system is used to record the voltage produced by the temperature transducer. After the data have been collected, each voltage value has to be converted into a corresponding temperature using this equation. Many data acquisition systems calculate the scale and offset from a pair of calibration values and automatically calibrate data after they have been collected.

The raw data recorded by the data acquisition system are in their most compact form (2 bytes per sample). Once the data are calibrated they are stored in floating point format, which typically occupies 5 bytes per sample. It is therefore common to save the data in uncalibrated form, but if this is done it is essential to ensure that the calibration coefficients (scale and offset) are recorded along with the data.

DATA ANALYSIS

Recording reliable data is half of the process of data acquisition and analysis. The second (and more interesting!) part is converting the raw data into useful numbers, a topic that is covered in Chapter 5.

chapter two

How a Data Acquisition System Works

A data acquisition system for a personal computer is made up from several parts. Firstly, there will be a circuit board (often called a *card*) covered with electronic components. This is the heart of the system and it contains several sections of circuitry, each of which performs a specific function (analogue-to-digital conversion, sample and hold, variable gain amplification and multiplexing). There is also some circuitry to allow the computer to talk to the analogue-to-digital converter, and some optional extras such as memory.

The circuit board usually plugs into a socket inside the computer. This socket is wired into the bus, the main 'information highway' of the computer. Specific installation instructions come with the board but you will need to turn the computer off and open the case. The sockets are soldered directly to the main circuit board of the computer. Depending on the type of computer, there may be a single socket or a row of them, and some may already have other circuit boards plugged in to them. There may also be more than one type of socket. For example, the standard bus format used in personal computers is the PCI bus, which is the one used by peripherals such as A/D boards and network cards. The computer may also have an advanced graphic socket, which is used for high-performance video cards and has a different shape compared to the PCI sockets. Older PC format computers used the AT bus for peripherals and pre-power PC Macintoshes used the now-obsolete Nubus format. New PC format computers often retain one or two AT style sockets so that they are compatible with older equipment. You can still buy A/D boards that use the AT bus but use PCI format boards if you can, because the performance is better and they can be used in both PC and Macintosh computers. All plug-in cards have a metal bracket on one end with connectors mounted on it. When the board is fitted into the computer the bracket lines up with a slot in the back of the case, so that the connector pokes through and allows cables to be plugged in when the case is closed. Standard peripheral boards provide the necessary circuitry to allow the computer to exchange information with a printer, a network, and so on. The data acquisition board does much the same thing: it allows the computer to gather information on the world around it by measuring temperature, pressure, light levels, and so forth. To measure temperature, for instance, a sensor is used that produces a voltage proportional to its temperature. The computer cannot measure this voltage directly and the purpose of the data acquisition board is to convert the voltage into a number, something the computer can work with. Thus a data acquisition circuit board acts as a 'translator' between the computer (which works with numbers) and transducers and other devices, which work with voltages.

Sometimes the circuit board does not plug into a socket inside the computer but is enclosed in a box that sits near to the computer. These external circuit boards also need a power supply, which may be inside the box or external. The external power supplies are usually the 'brick' type that plug directly into the wall socket and connect to the A/D board with a long cable (a wonderful term for these adapters is 'wall warts'!). There are two main reasons why external A/D boards are used. Firstly, they are less affected by the electrical noise from the computer, which is important when making very precise measurements. The Instrunet system from GW Instruments uses this approach. The second reason is that in some cases an internal board cannot be used. For example, the Macintosh iMac computer does not have any internal PCI sockets.

All external A/D boards have to connect to the computer using one of the external sockets on the back of the latter. There are A/D systems available that will fit on to virtually any of the sockets on the back of a PC or Macintosh but the performance varies considerably depending on which socket is used. The newest A/D systems use either the Universal Serial Bus (USB) or the Firewire (IEEE 1394) connector. Both USB and Firewire are fast connections that allow high-speed data acquisition, and they are compatible with both PC and Macintosh computers. The older SCSI connector allows medium-speed data acquisition but it is becoming difficult to find SCSI-based A/D systems. The slowest systems are those that use the serial or parallel printer ports. These systems are not of much use for general-purpose data acquisition but have a niche market for applications where a small, cheap data logger is needed.

Standard desktop computers are fine for data acquisition systems that are going to be used in the laboratory and are the cheapest way to buy computing power. If the system has to be portable, however, laptop computers are essential. The problem with laptops is finding space to fit the A/D board. One solution is to use an external A/D board that plugs into one of the ports on the back of the computer. The other solution, which is a lot neater and easier to carry around, is to use an

A/D board that fits into the PCMCIA socket. National Instruments makes a good range of A/D cards in PCMCIA format, which are compatible with both PC and Macintosh laptops.

The data acquisition circuit board acts as a translator between the computer and the world of transducers and voltages. The board must therefore be connected to both worlds. The connection between the computer and the data acquisition board is made when the board is plugged in or, in the case of boards outside the computer, via a short cable. A connection is also needed to the transducers on the 'voltage' side of the data acquisition board. Soldering wires from the transducers directly to the circuit board is very awkward and inflexible. The usual way is to use some sort of *break-out box*. This is a plastic or metal box with a row of terminals or connectors on it. Each terminal is connected to the data acquisition circuit board by its own piece of wire and all the wires are either laid side-by-side in a wide, flat ribbon cable or bundled together into a single thick cable. This cable goes between the break-out box and the data acquisition board, and is usually a few feet long so that the break-out box is not too close to the computer. Computers give off quite a lot of radiofrequency interference that can be picked up by sensitive amplifiers, and it is a good idea to keep transducer amplifiers a few feet away from the computer and monitor. At the computer end the cable goes through a slot in the back of the computer's case and then plugs into the circuit board.

Using a break-out box is a lot more convenient than connecting the transducers directly to the data acquisition board. Each transducer is connected to one socket or terminal on the break-out box so that eight or sixteen transducers may be connected to one box. The sockets make it easy to rearrange the connections for a different experiment. Using a break-out box also protects your computer and data acquisition board from damage if someone trips over the transducer cable.

Figure 2.1 shows the main parts of a data acquisition/analysis system, and the dotted line encloses the sections that are on the data acquisition circuit board. Each component of the system is now examined in detail.

ANALOGUE-TO-DIGITAL CONVERTER

The analogue-to-digital converter (ADC) is the heart of the data acquisition system. It converts a voltage applied to its input terminals into a number which can be read by the computer. Figure 2.2 shows what an ADC does. The ADC is designed to operate over some specific *range* of voltage, which is often 0 to 10 V or −5 V to +5 V. This range is divided into a number of evenly spaced levels separated by small, equal steps.[1] A common number of levels used is 4,096, which therefore have 4,095 steps between them. If the voltage range is 0 to 10 V then each step is

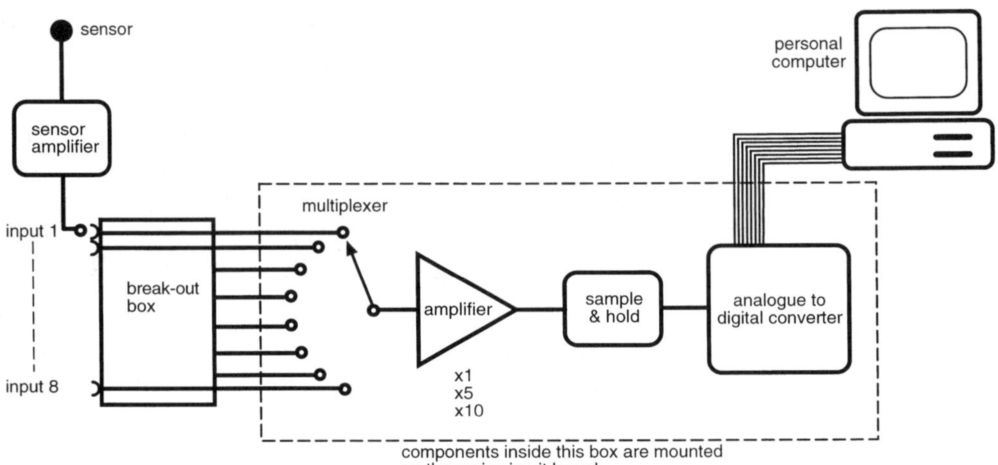

Figure 2.1 A typical data acquisition system broken down into functional blocks. The sensor and amplifier can be almost anything: a pH meter, a force transducer, a microelectrode, a mass spectrometer, and so forth. The various sensors are connected to the break-out box that is located close to the computer. The multiplexer, amplifier, sample-and-hold circuit, and the analogue-to-digital converter itself are electronic components mounted on a circuit board, which may be inside the computer or in a separate box. An important part of a data acquisition system that is not shown in this figure is the software which runs on the computer.

equivalent to 10/4,095 = 0.00244 V. The levels are numbered from 0 to 4,095. When a voltage is applied to the ADC, it compares it with the voltage corresponding to each level and the ADC output is the number of the level that most closely corresponds to the applied voltage. For example, if the voltage applied is 5 V then the output of the ADC will be 2,048, the number of the level that most closely corresponds to 5 V. This number is what is recorded by the computer when it 'reads' the ADC output.[2]

The vast majority of ADCs inside data acquisition systems use a method called *successive approximation* to convert the voltage into an equivalent number. For the technically minded a brief explanation of this process follows. Suppose that we apply 3 V to the input of the ADC we used in the example above: its range is 0 to 10 V and it has 4,096 levels with 4,095 steps between them. The ADC looks at the voltage applied to its input and decides if it is above or below the halfway point in its range. In our example the range is 0 to 10 V, so the halfway point is 5 V, and 3 V lies below this. The circuit then looks at just this bottom half of the range, 0 to 5 V, which includes the value of the input signal. It divides this new range in half and looks again to see if the applied voltage is above or below the new halfway point, which is 2.5 V. In our example the applied voltage, 3 V, is above the

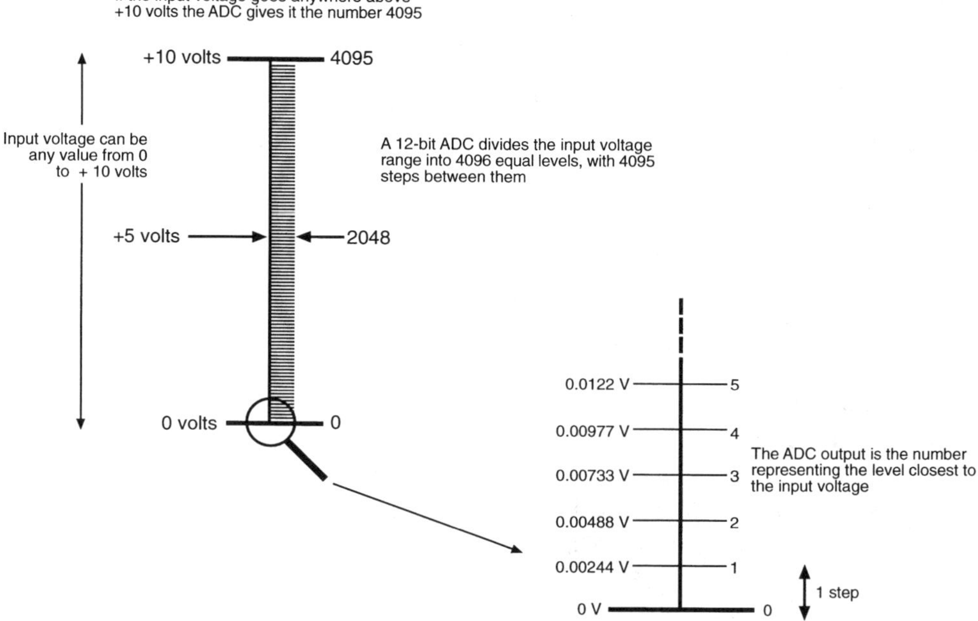

Figure 2.2 An ADC converts a voltage applied to its input into a number. The ADC divides its range into a number of levels separated by equal steps, and compares the applied voltage to this scale. Its output is the number of the level that corresponds most closely to the applied voltage. In this example the ADC has a range of 0 to +10 V that it divides into 4,096 levels with 4,095 steps between them. Each step therefore corresponds to $10/4,095 = 0.00244$ V.

halfway point. The circuit then looks just at this upper half of the range from 2.5 V to 5 V. Again, it divides this in half and sees whether the applied voltage is in the upper half or the lower half of the new range. The circuit keeps on doing this until the range it is looking at is equal to the step size, when it stops. It now looks at the answers to all those questions 'is the applied voltage in the upper half of the new range?'. It lines up all the answers, which for a 3 V signal applied to our 4,096 level ADC with a range of 0 to 10 V looks like:

no yes no no yes yes no no yes yes no yes

with the first answer on the left (12 answers in all). It is now very easy to convert this set of answers to the number 1,229, which represents 3 V. Just replace each 'no' with a zero and each 'yes' with a 1 to get the binary number 0100 1100 1101, which is 1,229 in base 10. In practice, the number is left in binary because this is the correct format for sending it to the computer.

The very fastest ADCs use another type of circuit called a flash converter. Flash converters are amazingly fast: for example, the AD770 flash converter made by Analog Devices only takes 5 ns (that is, the time it takes light to travel 1.7 m) to convert a voltage into its equivalent number. Flash converters have fewer levels over their range, 64 or 128 being typical. They are used for video image capture, radar and other specialized applications where their speed is needed and their low resolution is less important.

One obvious problem with dividing the input range into discrete levels is that the resolution of the ADC is limited to the step size. The output of the ADC is the number of the level that is closest to the input voltage – there are no 'half levels'. Increasing the number of steps will increase the resolution. For many years the standard laboratory ADC had 4,096 levels but increasingly people are buying 65,536-level ADCs as they become more affordable. The number of levels that an ADC has is usually specified by a number of 'bits' (*bit* is short for *binary digit*). An 8-bit ADC has 256 (2^8) levels, a 12-bit ADC has 4,096 (2^{12}) levels, and a 16-bit ADC has 65,536 (2^{16}) levels. This is because in the binary number system, which is what all computers use, 12 bits are needed to represent 4,096 individual numbers and so each time a 12-bit ADC converts a voltage to a number the computer that reads that number needs 12 bits of memory to store it.

Computers store numbers in multiples of 8 bits: a *byte* is equal to 8 bits. Programs which deal with numbers in multiples of 8 bits are fast and efficient. There is thus some advantage in making the number of bits of an ADC equal to 8 or a multiple of it because it is easier to deal with the data afterwards. The 8-bit ADCs need just one byte to store a sample but the resolution of 8-bit ADCs (256 steps) is too low for most applications. These ADCs are used when cost is more important than precision or when very fast sampling rates are needed. The next logical step would be 2-byte, or 16-bit ADCs. These used to be prohibitively expensive for general-purpose use and 12-bit ADCs were developed as a good price–performance compromise. They are fast, inexpensive and their resolution is good enough for most purposes. However, the cost of 16-bit ADCs has now fallen to the point where they are a good choice for new systems. Currently, 16-bit ADCs cost about twice as much as the 12-bit ADCs with comparable performance. However, slower 16-bit ADCs (20 kHz maximum sampling rate) that are fine for life-science work cost only about 50% more than a standard 12-bit ADC. In theory, each number from a 12-bit ADC only needs 1.5 bytes of storage space. In practice, they are almost always stored in 2 bytes, wasting 0.5 byte of memory per sample. This is because 2-byte numbers can be manipulated by computers much faster than 1.5-byte numbers and so some memory is wasted in return for faster processing. Also, the cost of computer memory and storage has fallen so much that it is not worth the programming effort needed to manipulate 1.5-byte numbers.

RANGE

It is important for accuracy that the range of the ADC matches the signal. Suppose we have an ADC with a range of -5 V to $+5$ V. If the signal from the sensor is small (e.g. 0.5 V maximum) then the step size is large relative to the signal and resolution will be poor (Figure 2.3). The step size is $10/4{,}095 = 0.00244$ V, which corresponds to $0.00244/0.5 \times 100\% = 0.48\%$ of the signal. If the input range is now reduced to -1 V to $+1$ V then the resolution is improved fivefold. On the other hand, we have to watch the signal level more closely because if the input goes higher than $+1$ V the ADC will still give the number 4,095 as its output, 'clipping' the signal. Similarly, if the input voltage goes below -1 V then the ADC will give the number 0 (Figure 2.4). If the signal goes way out of the input range then the ADC can be permanently damaged.

The best ADC range to use obviously depends on the size of the signal. The aim is to have the range of the ADC match the signal as closely as possible whilst avoiding clipping. Most laboratory equipment is designed to give an output of a few volts, and a common full-scale range is 0–5 V. For example, a pH meter might have a full-scale output of 0–5 V, which corresponds to a pH range of 0–14. In this

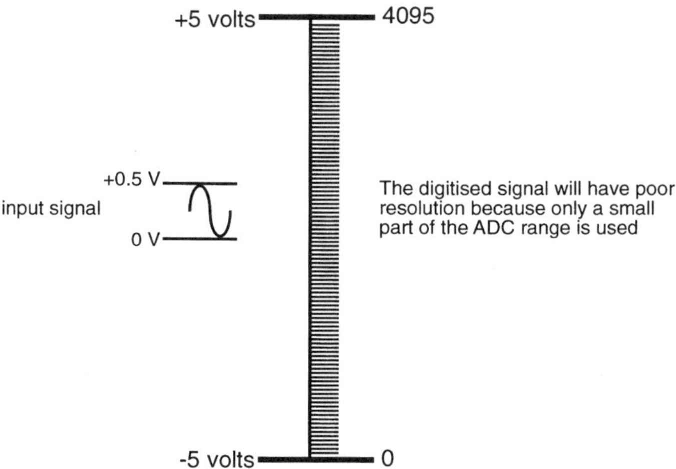

Figure 2.3 If the signal applied to the ADC is small compared to the ADC range the resolution of the digitized signal will be poor.

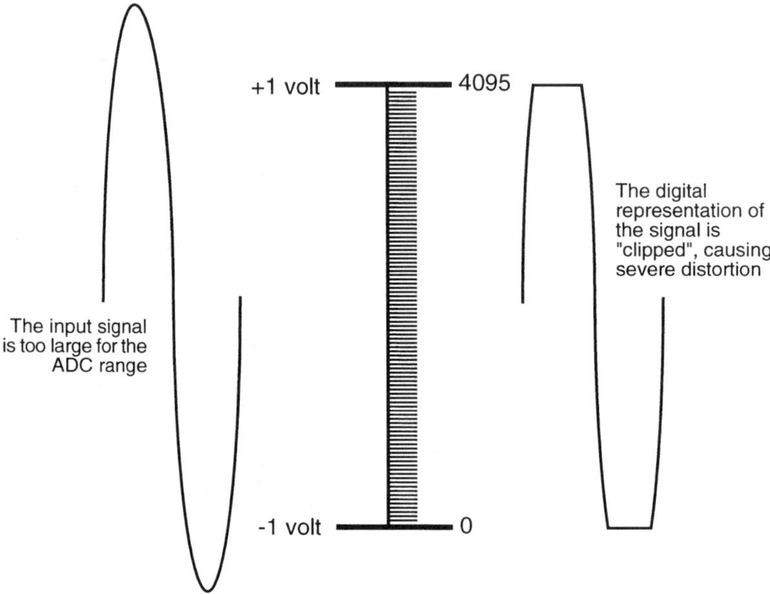

Figure 2.4 If the signal applied to the ADC is larger than the ADC range it will be truncated or 'clipped'.

case it is easy to match the range of the ADC to the instrument by setting the range to 0–5 V. If a 12-bit ADC is used then the resolution is $14/4{,}095 = 0.0034$ pH units, which is likely to be more than adequate in this case. If you do not know what the exact range of the signal will be, a useful rule of thumb is to make a guess at the expected range of the signal and set the ADC range to twice this or a little larger.

In some cases the ADC range is deliberately made smaller than the output of the equipment in order to improve the resolution. For example, most arterial blood pressure transducers work over the range -100 mmHg to $+300$ mmHg because pressures outside this range are physiologically impossible. Suppose the transducer amplifier converts this to a voltage in the range -5 V to $+5$ V, so that $+5$ V $= 300$ mmHg. If you want to record arterial pressures then setting the ADC range to the same as the amplifier, ± 5 V, will record any physiological pressure. The resolution is $600/4{,}095 = 0.15$ mmHg per step, fine for arterial pressures. Now suppose you want to use the same transducer and amplifier to record central venous pressure. The maximum range of the latter is about ± 10 cmH$_2$O, or ± 7.36 mmHg. A resolution of 0.15 mmHg (0.2 cmH$_2$O) is now inadequate and a smaller range on the ADC is required. The amplifier is still the same, so 7.36 mmHg corresponds to $7.36/300 \times 5 = 0.13$ V. We should set our ADC to the next largest pre-set range, usually ± 0.5 V.

How is the ADC range changed? Take a look at Figure 2.1. The ADC converter itself, which performs the conversion of the signal into a number, is a single integrated circuit (*IC*). The input range of this integrated circuit is limited to only one or two values, usually ± 5 V and 0–10 V. To make the whole data acquisition system more versatile, the manufacturer puts an amplifier in front of the ADC. This amplifier has a number of pre-set *gains* such as $\times 1$, $\times 5$, $\times 10$, $\times 50$, and $\times 100$. The gain is the factor by which the amplifier increases the size of a signal applied to its input. For example, if the gain is set to $\times 10$ then a signal applied to the amplifier will be increased in size (*amplified*) 10 times, so a voltage of -0.5 V applied to the input of the amplifier will be increased to -5 V. This amplified signal is fed into the ADC and thus the effective range of the ADC/amplifier combination is much wider. So, if the basic range of the ADC is ± 5 V then with an amplifier gain of $\times 1$ the range of the ADC/amplifier combination is also ± 5 V. If the gain of the amplifier is now changed to $\times 10$, the range of the ADC/amplifier combination is ± 0.5 V. In this way the range of the system can be altered to match the signals applied to the input.

The gain of the amplifier, and thus the range of the system, is set by a switch. On some units this is a mechanical switch soldered to the A/D board that you have to change by hand. On others it is an electronic switch set by the computer. Electronic switches have the advantage that they can be changed much more rapidly than a mechanical switch. Usually, the range has to be the same for all the inputs, because they all go through the same amplifier, but the more advanced data collection systems allow you to program different gains for each input. In this situation the computer changes the gain of the amplifier before taking each sample. There is, however, a limit to how quickly the gain can be changed if accuracy is to be preserved. In practice, most people leave the gain fixed at one value for all inputs during data collection and only change the gain when the experimental setup is altered.

The *settling time* is often quoted in the specifications of a data acquisition system. This is the time taken for the output of the amplifier to catch up with a change in the voltage applied to its input. The settling time is usually quoted as the time taken for the amplifier's output to settle to a certain percentage of its final value after change in the input voltage. For example, the settling time may be quoted as '5 µs to settle to 95% of final value'. This is important because the output of the amplifier is connected to the ADC and if the output of the amplifier has not settled before the ADC starts to read it the value will be inaccurate. Obviously, the manufacturers ought to make sure that the amplifier settles quickly enough so that even at the highest sampling rate the amplifier output is an accurate reflection of its input. This is not always the case and if you are sampling at the maximum rate and need full accuracy you should check that the amplifier can keep up with the ADC. The settling time of the amplifier is usually worse at higher gains so for the best accuracy at high sampling rates use modest gains ($\times 1$ to $\times 10$).

UNIPOLAR AND BIPOLAR MODE

Data acquisition systems usually have two modes of operation called *unipolar* and *bipolar*. This refers to the polarity of the signal that the ADC will accept at its input. In unipolar mode the ADC will only give meaningful values if the voltage applied to the input is positive. Negative voltages, whatever their value, are given the same number as $0\,V$. The gain, and thus the range, of the data acquisition system can still vary, so typical ranges in unipolar mode are 0 to $+1\,V$, 0 to $+5\,V$, and 0 to $+10\,V$. Small negative voltages applied to the input will not damage the system but will be given the same number as $0\,V$.

In *bipolar* mode the data acquisition system will accept both positive and negative voltages. The range is always symmetrical about $0\,V$, that is, a range of $-2\,V$ to $+5\,V$ is not possible. Typical ranges are $-1\,V$ to $+1\,V$, $-5\,V$ to $+5\,V$, and $-10\,V$ to $+10\,V$. These are often abbreviated to $\pm 1\,V$, $\pm 5\,V$, and $\pm 10\,V$, respectively.

Changing between unipolar and bipolar mode is done either by setting a switch on the data acquisition board itself or via the computer. The former method is more common. The switch is located on the circuit board and so changing it requires taking the computer apart – not something you want to do every day, and most people set the mode once and leave it. In some cases a *jumper* is used to set the mode instead of a switch. A jumper is a small connector that completes an electrical circuit between two metal contacts. The metal contacts are small pins about 6 mm long that stick up from the surface of the board. The jumper is pushed over a pair of pins, connecting them together. This alters the circuitry so that the ADC works in the correct mode.

Deciding on which mode to use (unipolar or bipolar) is fairly easy:

   i. If you want to measure negative voltages use bipolar mode.
   ii. If you are not sure what sort of signals you will be measuring, or you use your data acquisition system for a number of different experiments, set it to bipolar mode.
   iii. If you are certain that the signals from your sensors will not be negative, and you need the best possible accuracy, you can use unipolar mode.
   iv. If in doubt use bipolar mode.

You may wonder why anyone uses unipolar mode. The reason is that if you only deal with positive voltages the resolution is twice as good in unipolar mode as in bipolar mode. For example, suppose you want to record blood pressure invasively using an arterial catheter and a pressure transducer. The amplifiers for blood pressure transducers typically have an output voltage of 1 V per 100 mmHg. The arterial blood pressure will never be negative so neither will the signal from the transducer amplifier. The range of most blood pressure transducers is -100 mmHg to $+300$ mmHg, so the maximum signal coming from the amplifier will be $+3$ V (corresponding to a blood pressure of 300 mmHg). The range of the ADC therefore has to be at least 0 to $+3$ V to avoid clipping. The nearest range is 0 to $+5$ V, so the ADC can be set to unipolar mode and the range to 0 to $+5$ V. The resolution with this range and a 12-bit ADC is 5/4,095 V $= 1.22$ mV (a 12-bit ADC divides its range into 4,095 steps, and the resolution is equal to one step). This corresponds to a resolution of 0.12 mmHg (1.22 mV $\times$ 100 mmHg/ V $= 0.12$ mmHg).

If bipolar mode is used, a range of -5 V to $+5$ V is needed in order to accommodate a signal of $+3$ V without clipping. This is a total range of $5 - (-5) = 10$ V. The resolution is therefore $10/4{,}095 = 2.44$ mV because the 4,095 steps are spread out evenly between -5 V and $+5$ V. In terms of pressure the resolution is 0.24 mmHg, which is half the resolution obtained when we restricted the range to 0 to $+5$ V.

MULTIPLEXING

In most cases we need to record more than one signal at the same time. Several ADC chips could be used but there is a cheaper alternative, called *multiplexing*. Suppose we need to record eight signals at once (Figure 2.1). A single ADC and its associated amplifier are connected to each signal in turn through an electronic switch called a *multiplexer*. The ADC measures the first signal, the computer reads it and then the multiplexer switches the amplifier to the second signal. Once all eight signals have been read, the multiplexer switches back to the first signal and the process starts over again. The computer keeps track of which number corresponds to which signal. Each connection to the multiplexer is called a *channel*; the system shown in Figure 2.1 is therefore an eight-channel data acquisition system.

One problem with multiplexing is that the ADC has to be much faster than if one ADC per channel is used. Suppose we have an eight-channel system and we need to record the value of each signal 100 times a second. Since there are eight channels the ADC has to be able to convert at least 800 samples per second. This is

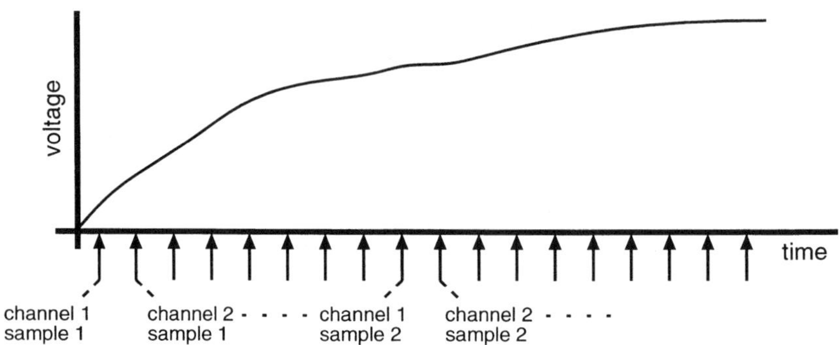

Figure 2.5 Sampling skew due to multiplexing. Samples recorded from adjacent channels are not simultaneous and this must be taken into account in subsequent calculations.

something to watch for in the fine print of advertisements for data acquisition systems. A system may be advertised as '8 channel, 100 kHz sampling rate'. This sampling rate usually only applies when only one channel is being used, and the maximum sampling rate falls to 12.5 kHz (100/8) when all eight channels are used.

There is also a more subtle problem with multiplexing called *sampling skew*. Because the samples are taken sequentially there is a delay between each one: they are not simultaneous. In many cases this is not important but it can lead to errors in some circumstances. Figure 2.5 shows what happens when a signal is multiplexed. In this example the same signal is connected to all eight inputs of the data acquisition system. The multiplexer connects the eight channels one at a time to the ADC, so that channel 2 is connected slightly later than channel 1 and the input signal has changed a little. After all eight samples have been taken, the multiplexer starts with channel 1 again. When the data are presented on the computer screen, they will look as if all the successive sample numbers (sample 1, sample 2, etc.) were taken at the same time for each channel whereas in fact they are all slightly different. This difference can be important if you are interested in the relative timing between channels. For example, channel 1 might correspond to a brief flash of light and channel 2 the resultant surface EEG from the occipital area of the head.[3] The latency of the EEG is the time delay between the two signals, but there will be an additional delay between the two signals due to the multiplexer that we must take into account when calculating the latency. The multiplexer-induced delay can usually be estimated from the sampling rate and the number of channels used because the multiplexer normally spaces the samples out evenly. For example, suppose the sampling rate is 100 Hz and recordings are being made from eight channels. Successive samples on each channel are 10 ms apart, but because there are eight channels the ADC is actually sampling at 800 Hz and the delay between adjacent channels is 1.25 ms. If the optical stimulus was recorded on channel 1 and the EEG on channel 2 then 1.25 ms would need to be subtracted from the apparent latency to obtain a true value.

Sampling skew can be reduced in several ways. The easiest is to sample at a faster rate but the amount of data recorded can become large. Sometimes the ADC can be set up to record from all the channels sequentially as quickly as it will run, then wait until the next sample is due. Suppose the sampling rate is again 100 Hz and eight channels are in use. The maximum sampling rate of the ADC is much higher, 50 kHz for example. The ADC is set up so that it records from all eight channels sequentially at the fast rate (50 kHz). The delay between successive channels is then 20 µs. When all eight channels have been sampled, the ADC waits the remaining 9.86 ms until the next round of samples is due.

A more expensive method that effectively eliminates sampling skew is to use a sample-and-hold circuit (see the next section) on each channel. When it is time to record a set of samples the signals on all channels are effectively 'frozen' by the sample-and-hold circuit. Now the ADC can convert each 'frozen' channel at its leisure with no sampling skew.

SAMPLE-AND-HOLD CIRCUIT

The maximum sampling rate of an ADC is set by its *conversion time*. This is the time it takes the ADC to convert a voltage applied to its input to a number. When the ADC is going as fast as it can, it will start converting the next sample as soon as the previous conversion is completed, so individual samples will be spaced one conversion time apart. The maximum sampling rate of an ADC is thus 1/(conversion time). For example, an ADC with a conversion time of 10 µs has a maximum sampling rate of 100 kHz. In the conversion period the ADC is working out the number that most closely corresponds to the voltage applied to its input. During this time the voltage in the ADC input must not change much or the number produced by the ADC will not be an accurate reflection of the voltage. To make sure the voltage does not change, a *sample-and-hold* circuit is usually built into the system. This circuit 'freezes' the input voltage when the ADC starts converting and keeps it constant until conversion is complete. It is a bit like a sketch artist trying to draw a moving scene such as a busy street or an animal playing. It takes the artist about 5 min to sketch the scene, and of course the scene changes in this time period. To get around the problem the artist takes an instant Polaroid photograph of the scene and draws from that instead. The photograph 'freezes' the scene so that an accurate sketch of what it was like at that time can be made. When one sketch is finished, another photograph is taken and sketched. In the same way the sample-and-hold circuit takes a 'snapshot' of the voltage and keeps it still so the ADC can convert it accurately.

SINGLE-ENDED AND DIFFERENTIAL INPUTS

All ADCs measure voltages. In order to measure other variables such as temperature, a transducer and extra circuitry are used to convert the temperature into a representative voltage (see Chapter 4). It is important to appreciate that voltages are always measured between two points. For example, a '12 volt' car battery produces a voltage difference between the two terminals of 12 V. When using a meter to measure voltages, two leads are always connected from the meter to the object being tested. The meter displays the voltage difference between the two leads. The two leads on a voltmeter are labelled + and −; the + lead is red and the − lead is black. The meter is made so that if the red lead is more positive than the black one the meter gives a positive answer. For example, a car battery's + terminal is 12 V more positive than its − terminal. If the red lead of a meter is connected to the + terminal and the black lead to the − terminal, the meter will read +12 V, because the red lead is 12 V more positive than the black one. What happens if the connections are reversed so that the red lead is connected to the − terminal of the battery and the black lead is connected to the + terminal of the battery? The red lead is now 12 V more negative than the black lead so the meter reads −12 V.

An ADC is a voltmeter so it has two connections and it measures the voltage difference between them. A typical data acquisition system has eight channels. Each channel measures a voltage and so must have two connections or *inputs*, which gives a total of sixteen connections. The two inputs are labeled + and −, the same as a voltmeter. Very commonly, one input from each channel (the − input) is joined to the same point on the circuit board and this common point is called the *analogue ground* or *earth* connection. Now we have eight individual connections to the eight channels and a ninth ground connection which is joined to the − input on all the channels, as shown in Figure 2.6 (for clarity only three channels are drawn). An ADC wired in this way, with one connection from each input connected to ground, is said to be wired in *single-ended* mode. Each channel appears to have a single connection, the other one is shared amongst all the channels. If we want channel 1 to measure a particular voltage, we need to apply this voltage across the two connections to channel 1. One of these is the + input that is unique to channel 1, the other is the ground connection that is shared with all the other channels. Voltages measured in this way are described as 'relative to ground',

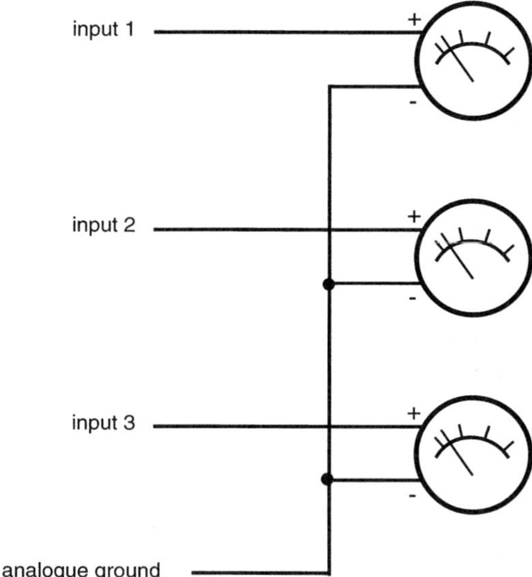

Figure 2.6 Three data acquisition channels wired in single-ended mode. The channels are represented by voltmeters. Each channel has one connection unique to it, and there is a ground connection which is common to all channels.

which emphasizes the fact that the voltage difference between the + input and the ground connection is being measured. A big advantage of joining one connection from each channel to ground is that the total number of wires needed is almost halved, from sixteen to nine.

In some cases single-ended measurements are not ideal. Figure 2.7a shows an amplified signal from some sensor being sent down a long cable to an ADC wired in single-ended mode. This long cable is likely to pick up some noise from power lines and other sources, which will distort the signal and introduce errors. It is also not that uncommon for one piece of equipment connected to an ADC to generate unwanted signals on the ground connection, and these unwanted signals will affect all the channels if they are wired in single-ended mode.

One solution to the problem is to wire the ADC channels in *differential* mode. Figure 2.7b shows two channels of an ADC connected in differential mode, with a piece of equipment (a sensor amplifier) wired up to channel 1. As before, the ADC measures the voltage difference between the two inputs to each channel. What is different is that the inputs from each channel are kept separate rather than one from each channel being connected to ground. There is still a ground connection but the ADC ignores it completely because the inputs are not joined to it; what the ADC measures is the voltage difference between the two input connections. These two input connections are called + and −, as before. If the + input is more positive than the − input then the ADC will call that voltage positive and, conversely, if the + lead is more negative than the − lead then the ADC will call the voltage

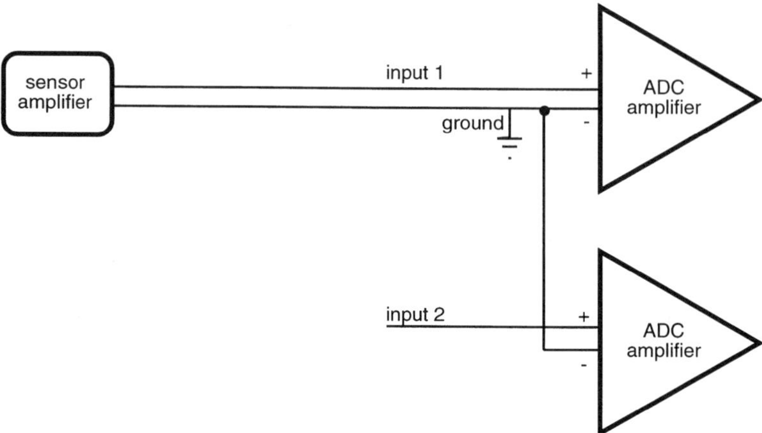

Figure 2.7a Single-ended amplifiers are susceptible to noise picked up by long cables and electrical noise on the ground line.

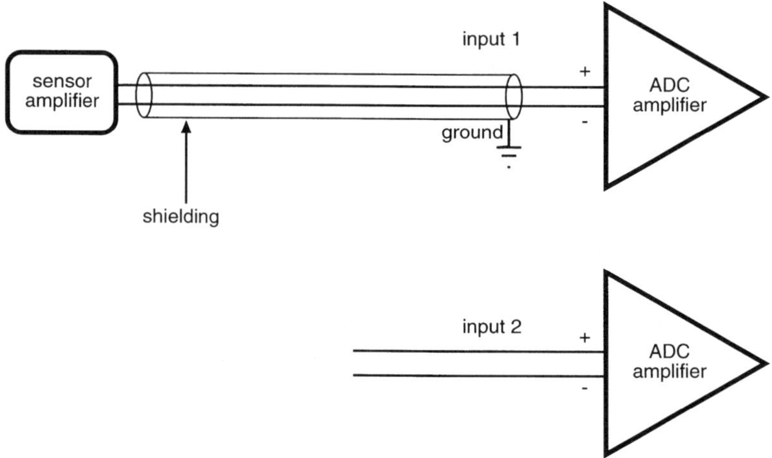

Figure 2.7b Amplifiers connected in differential mode are able to reject noise picked up by long cables, and channels are more isolated from each other.

negative. Using an ADC in differential mode puts a bit more responsibility on the user to wire it up correctly so that the polarity of the reading given by the ADC is correct. In single-ended mode the ground is always connected to the − input so mistakes are unlikely.

The advantage of a differential input is that the noise picked up by long cables causes smaller errors in measurements. The extra noise tends to be the same in both wires and, because the amplifier measures the difference in voltage between the two wires, the noise does not affect the measurement. Signals or noise that are the same in both inputs of a differential amplifier (relative to ground) are called *common-mode* signals. The whole purpose of a differential amplifier is to amplify differential-mode signals and reject common-mode signals. How well the amplifier does this is measured by the *common-mode rejection ratio* (CMRR). The higher the CMRR of an amplifier, the better it will reject common-mode noise. If you have to use long cables you can improve things further by using a shielded cable between the sensor amplifier and the ADC and connecting the screen to ground (Figure 2.7b). In fact, whatever the ADC setup (single-ended or differential), there always has to be a ground connection between the ADC and the sensor amplifier. In differential mode this ground connection does not carry any information, it is just there to make sure that big voltage differences (due to static electricity, for example) do not build up between the ADC and the sensor.

Differential amplifiers are often used when noise is a problem. A common application is the ECG amplifier, which has to measure a small signal of 2–3 mV generated by the heart and ignore all the 60 Hz noise from the wiring in the room. It achieves this by measuring only the difference in voltage between the active electrodes that are placed across the heart. The 60 Hz signal from the wiring is the same in both electrodes (i.e. it is a common-mode signal) and, because the amplifier has a high CMRR, it rejects this interference. On a practical note, the ultimate CMRR of any system that uses a differential amplifier is set not only by the CMRR of the amplifier but also by how well the two amplifier inputs are balanced. For the highest CMRR both amplifier inputs should be treated in exactly the same way – for example, the same-length cables and same electrodes should be used for both inputs. For other ideas on how to reduce the 60 Hz interference, see Chapter 5.

Data acquisition/analysis packages usually give you the option of either single-ended or differential mode. There is a catch – the maximum number of channels available is usually half as many in differential mode as in single-ended mode. Changing from single-ended to differential mode or vice versa involves rewiring the connections to the ADC and is not something you want to be doing every day. For the majority of applications single-ended mode is fine and the wiring is simpler.

A final note on ADC inputs: there are usually two grounds on the manufacturer's wiring diagram, *analogue ground* and *digital ground*. There is a good reason for having two ground connections. The circuits on the data acquisition board that talk to the computer take quite a lot of power and, if they use the same ground connection as the inputs, they can add noise to the latter. A well-designed analogue to digital circuit board therefore has two grounds, analogue and digital. The analogue ground is used for the inputs and the digital one for outputs such as the interface to the computer and digital input/output connections (see Figure 2.10). One can think of the analogue ground as being delicate and precise: it will measure small voltages accurately but cannot cope with abuse such as high currents. The digital ground is rough, noisy, and powerful: if you do not mind a bit of noise you can connect almost anything to digital ground and it will cope with it.

COMPUTERS

The vast majority of computer-based data acquisition systems use either PC type or Macintosh personal computers. Large shared mainframe or workstation computers are quite useless for this sort of work for two main reasons:

    i. they cannot be moved around the laboratory from one experiment to another;

    ii. data logging requires precise timing and the computer cannot be interrupted when it is collecting data or it will miss samples.

Workstations and mainframes are designed to be shared and this is incompatible with the need for uninterrupted use by one person. If you have ever worked on a network or similar shared computer system you will know how slow the response can be when there are a lot of people using the system. Imagine trying to do something that requires accurate timing on such a system: it just cannot be done unless you are the only person using it.

The majority of data acquisition systems are based on PC-compatible computers and there is a greater choice of hardware and software available than for Macintoshes. However, the current leading manufacturer of data acquisition systems (National Instruments) supports both platforms and continues to bring out upgrades for both the PC and the Macintosh. If you have an investment in Macintosh hardware, or (like the author!) just prefer working with them, you can still build excellent Mac-based data acquisition and analysis systems. One welcome trend is that the interface connections on personal computers are becoming standardized cross-platform so that manufacturers no longer have to make two different types of hardware. The different connection formats (PCI, USB, Firewire, etc.) were explained in detail earlier in this chapter.

SOFTWARE

So far we have concentrated on the 'hardware', the electronic circuitry that is needed to translate a voltage from the real world into a number that a computer can understand. On its own, the hardware cannot do very much. It needs instructions from the computer to tell it which channel to use, what range to set, and so on. This is performed by a simple program called a *driver* program which sets up the number of channels, gain, sampling rate, and so forth, and then records a short block of data. Driver programs are tricky to write because they require a detailed knowledge of the computer's architecture in order to keep the timing between samples accurate. Luckily, they are always supplied with the hardware. Drivers are designed to be incorporated into larger programs and on their own they are not of much use.

Many applications for data acquisition systems require only that a signal be recorded so that it can be viewed at a later date. In this sense the data acquisition system is acting like the traditional FM tape recorders or chart recorders that have been used in laboratories for years. Almost all ADC systems include software that allows them to emulate these functions – data can be recorded, saved to disc, viewed on the screen, and measurements made using some form of cursor. The system may also record the measurements in a format that can be directly imported into a spreadsheet or graph-drawing program. Many inexpensive hardware/software options have this level of functionality, which is basically a user interface for the A/D hardware. What these packages do not do is perform any analysis on data. There is a world of difference between recording data with a data acquisition system and obtaining useful numbers from them. The raw data are not very useful: what one needs are parameters that describe the data in a compact form. For example, suppose that an ECG signal has been recorded from a subject at 1,000 samples/s in a study that looks at changes in heart rate with exercise. The raw data consist of measurements of the instantaneous ECG voltage made every 1.0 ms. These measurements are not really very useful: what is needed is the heart rate, calculated as the mean value over the last ten beats. Obtaining the heart rate from raw ECG data is not a simple task. The R wave in each ECG complex must be identified, the time between successive R waves measured, converted to a rate and then averaged over the last ten beats. The program needs to be able to pick out the R wave reliably in the presence of extraneous noise such as respiratory artefacts, EMG signals from muscle activity (a big problem in exercise studies), and the 60 Hz interference. Basic data acquisition software does not do this. What is needed are some powerful tools to analyse raw data and turn them into meaningful numbers. This is where the advanced analysis programs are so useful.

The most complete data acquisition/analysis software is LabView from National Instruments. This software provides all the functions needed to build a complete system; control of the A/D converter, data collection and storage, user interface controls, graphical displays, and a good range of advanced signal processing algorithms. However, this functionality comes with a steep learning curve that is daunting and requires a considerable investment of time to master. National Instruments realized that not everyone was able to make this investment and brought out a range of ready-to-use software packages. One of these, BioBench, is specifically aimed at the life sciences field. BioBench has a functionality similar to other software designed for physiological recording: data can be recorded, played back, and stored but there is limited analysis capability.

If advanced data analysis is needed then there are a couple of ways to build the system. The easiest route is to use a fairly simple program that collects the data and provides a friendly user interface. The block of data is then processed by a second set of programs that do the serious number crunching. There are some good packages available for advanced data analysis such as Matlab, IDL, HiQ, and Mathematica. As a last resort, data analysis routines can be written in a low-level language such as BASIC, Pascal, or C/C^{++} but this is a slow job even for a good programmer. User-written programs that have been compiled can often be called from within a high-level program so that only the most critical functions have to be written in the low-level language. A second approach is to write the entire program in LabView (National Instruments), which has a good range of advanced signal processing functions. User-written compiled code can also be linked into the program. A third approach is to write the program in a low-level language such as C/C^{++} but making extensive use of library functions, saving the C code for 'glue' and critical areas. This is the idea behind National Instruments' LabWindows/CVI software.

Tasks for data acquisition/analysis systems are often classified into *real-time* and *offline* depending on whether calculations are performed at the same time that data are being collected ('real-time') or at some later date ('offline'). Many laboratory experiments are offline. During the experiment the data acquisition/analysis system is just used to collect and store data and the calculations that give the final results are performed after the experiment is over: for complex problems this may take weeks of work. Contrast this with a system used to monitor the blood pressure of an anaesthetized patient. In this application it is obviously essential that the results are available to the anaesthetist straight away. So at the same time as the data acquisition/analysis system is recording the blood pressure from a transducer it has to calculate the mean, maximum (systolic), and minimum (diastolic) pressure for every heartbeat, average them over the last five beats and display the result along with the pressure signal from the transducer on a monitor.

Increasingly, real-time data analysis is used in laboratory experiments. The benefits are clear: the results of a procedure are available immediately to the user, who can choose to continue the experiment or repeat the procedure if the data do not look good. It eliminates the hours that used to be spent organizing and processing data at the end of each experiment, which is crucial in an industrial environment.

One problem of real-time data collection and analysis is that it puts heavy demands upon the host computer. Luckily, the speed and power of personal computers has increased enormously in the last few years, which has made the development of complex real-time systems feasible. There are also standard techniques that take some of the load off the computer and transfer it to the A/D hardware. One of these frees up the computer from the task of fetching and storing each sample from the ADC when it is ready. Instead, the data are held in a temporary electronic store called a *buffer*. A buffer is a circuit containing some memory chips and it is placed in between the ADC and the computer. In some cases it is physically located on the ADC board – a 'buffered ADC board'. The ADC and buffer circuits collect data on their own without any intervention from the computer until the buffer is full; the computer then copies all the numbers from the buffer into its own storage system in one go. Moving a block of data is much more efficient than moving samples one at a time and while the buffer is filling the computer can carry on with data analysis. Buffers are also useful when collecting large amounts of data rapidly because the buffer circuit can collect samples much faster than the computer.

As sampling rates increase, the small amount of time taken for the computer to empty the buffer is still too long, which leads to gaps in data collection. Such gaps are intolerable in many applications where continuous or *seamless* data collection is essential. The solution is to use two buffers (*double buffering*). When the first buffer is full, the ADC switches immediately to the second one without missing a sample; the computer now can empty the first buffer whilst the second one is filling. As long as the computer empties the first buffer before the second one is full, there will be no breaks in data collection. Double buffering is essential in complex, real-time data analysis systems because it maximizes the amount of time that the computer has to process the data and update the displays.

The second technique for handling large amounts of data at high sampling speeds is to use *direct memory access* (DMA). With DMA, the ADC board stores the samples directly in the computer's main memory area (RAM) without any intervention from the computer. This technique is often combined with buffering because it allows the A/D card to use the host computer's RAM as a buffer rather than having to have memory chips on the A/D card itself. This lowers the cost of the A/D card.

As an aside, the term 'megabytes' is used a lot in computing. In strict scientific terms, one megabyte (usually abbreviated to 1 Mbyte) is 1,000,000 bytes because the prefix 'mega' means 'multiply by one million'. Similarly, one kilobyte in strict scientific terms is 1,000 bytes. Memory circuits used in computers always contain a number of bytes that is a power of two because this is the most efficient way to organize the memory. It so happens that some powers of two are close to powers of 10 as well. For example, $2^{10} = 1,024$ which is very close to 1,000 (10^3). Well, fairly quickly computer scientists became tired of talking about '1,024 bytes' and it was abbreviated to '1 kilobyte', even though the two are not the same. Computer scientists knew that 1 kilobyte was really 1,024 bytes because they never talked about anything else. Using the same logic, $2^{20} = 1,048,576$, which was close enough to 1,000,000 that no-one really minded. So the term 'megabyte' has come to mean 1,048,576 bytes even though this is not strictly correct.

DATA FORMAT AND STORAGE SPACE

Once data have been recorded they nearly always need to be manipulated mathematically. As an example, suppose that you wanted to record the velocity of air at the mouth of a person who is exercising on a bicycle. One way of recording air velocity is with a *Pitot tube*. A Pitot tube is a pair of thin metal tubes that are placed in the airstream, one facing upstream and one facing downstream. The key feature of a Pitot tube is that the pressure difference between the two tubes is related to the airflow velocity. The pressure difference is easily measured with a pressure transducer, which generates a voltage proportional to the pressure difference between the tubes. It is this voltage that is recorded with a data acquisition system.

From the raw voltage data we need to work out the airflow velocity. The voltage from the pressure transducer and the airflow velocity are related by the following equation:

$$\text{velocity (m s}^{-1}) = k \times \sqrt{\text{voltage (V)}}$$

k is a calibration factor that is either measured or taken from the Pitot tube manufacturer's data sheets.

In order to convert each data point from V to m s^{-1}, two calculations must be performed:

    i. take the square root,
    ii. multiply by k.

Obviously, these calculations should be made as accurately as possible. To ensure that this is so, all of the advanced data analysis programs use *floating-*

point format. This is a way of representing a number inside a computer to a high degree of accuracy. The details of how this is done are not important except for one thing: it takes between 5 and 10 bytes of computer memory to store each number in floating-point format.[4] On the other hand, it only takes 2 bytes of memory to store a sample from a 12-bit ADC. Thus as soon as we take our raw data from the ADC and do some manipulations the amount of memory needed for storage goes up several fold.

For small amounts of data this is not a big problem but it is important to consider it if a lot of samples are going to be taken. Suppose that in this study we are interested in the airflow patterns at exercise. A high sampling rate will be needed because the complex airflow patterns change rapidly, and we have worked out that a sampling rate of 200 Hz is required (Chapter 3 explains in detail how to work out the correct sampling rate). The study takes 30 min to complete. At the end of the study there will be $30 \times 60 \times 200 = 360,000$ samples. Each sample takes up 2 bytes, so these take up 720,000 bytes of storage space. This is not too bad: the samples can be stored in the computer's memory while they are being collected and at the end of the experiment the data will fit on one floppy disc for archiving.

Now convert the raw data into airflow velocities, using the formula above. The program will convert the numbers to floating-point format and thus each calculated value takes up at least 5 bytes. This will require 1,800,000 bytes to store all the data, more than the capacity of a floppy disc. There are ways round the problem (such as using a Zip disc to store the data) but the message is clear: if you intend to collect large amounts of data do some planning to work out how much space they will occupy and how you are going to store the data for safe-keeping. It is clearly more efficient to store data in its raw form than in floating-point format.

Data acquisition systems can store raw data in one of two places: in the computer's main memory (RAM) or on the hard disc. The most convenient place is in RAM because it is very fast and available for analysis. When all the data have been collected they are transferred to the hard disc for storage (all the information in RAM is lost when the program is closed or the computer is turned off, so data need to be saved to disc as soon as they have been collected). For measurements that last only a few minutes, this is the most convenient way to handle the data. It is also the way to go if very high sampling rates (in the Mbyte/s range) are used since the computer cannot save the data to disc quickly enough. In the life sciences these high sampling rates are not needed because biological systems do not change that quickly. Life science experiments also often involve long periods of data collection. Under these conditions the data can be saved to disc as they are collected. A double-buffered A/D converter is used with the buffer size typically set to a few kilobytes. Each time the buffer fills up, the computer saves the block of data to disc, appending it to the data already there. The length of data storage is limited only by the size of the hard disc and the capacities of the latter have increased enormously in recent years so that a typical mid-range computer now comes with a hard disc capacity of 8–12 Gbyte, and 20 Gbyte hard discs are readily available. A data acquisition system collecting data from 16 channels at 200 samples/s and storing it in floating-point format (5 bytes/sample) generates 16,000 bytes of data per second, or about 1 Mbyte/min. A 10 Gbyte hard disc will hold nearly seven days worth of data, which is far more than most applications will need.

At the end of the experiment two copies of the data (one master, one backup) should be made and the space freed up on the hard disc for the next experiment. It is not advisable to keep data on a hard disc, it is too vulnerable to disc crashes and accidental erasure. Storing the raw data can be a real problem. The data may have to be archived for many years to satisfy government or institutional guidelines. Floppy discs hold 1.4 Mbyte of data so they are fine for small bits of data but unwieldy for large amounts. Blocks of data larger than 1.4 Mbyte can be split amongst several floppy discs using an archiving program such as Stuffit™ (Aladdin Systems, Inc.) but once the data need more than about five floppy discs for storage the files become unwieldy to use. If you plan to collect large amounts of data you will need to use some other mass storage system. There is a small revolution going on in mass storage systems at the moment and prices are falling rapidly. The traditional system is a tape streamer. The tapes are cheap and have a high capacity but the slow access time makes them tedious to use, and they are best used for archiving only. The two big growth areas are removable hard drives and optical drives. Removable hard drives such as the Zip and Jaz from Iomega are very cost-effective for mass storage and they are fast. The Zip format is widely used and many computers come with Zip drives as standard. Magneto-optical drives used to be popular but are being taken over by compact disc technology, and both write-only (CD-W) and read–write (CD-RW) systems are available. Compact-disc systems have the advantages that the data can be read by any computer with a CD-ROM drive and the blank discs are extremely cheap. Another advantage is that the stored data are very safe: the discs are immune to magnetic fields and the lifetime of the data is much longer than magnetic media. Data stored on magnetic media gradually 'fade'. It is difficult to say how quickly this happens but it is generally accepted that data should not be stored for more than five years on magnetic media without being 'refreshed' by rewriting them. On the downside, writing to CD discs is slow and not as straightforward as magnetic media because the data must conform to CD-ROM format. Having said all this, one of the biggest problems with data storage is the rapid technological progress which means that discs become unreadable within 2–3 years because the drives are no longer available.

The size of data files can be reduced before storage by using a data compression program to compress them. The size of a compressed file is usually about 60% of the original so that a 1.4 Mbyte floppy disc full of compressed data is equivalent to 2.3 Mbyte of uncompressed data. It is important to realize that there are two types of compression programs: *lossless* and *lossy*. With a lossless compression program you get back the original file when it is uncompressed. No information is lost. In contrast, lossy compression programs lose some data when the file is compressed so you do not get back quite the same information that you saved. Obviously, only lossless compression programs should be used for storing scientific data. Examples of lossless compression programs are PKZIP^{TM} and Stuffit^{TM}. Lossy compression algorithms have much higher compression ratios (up to 100:1) and they are mainly used to compress video pictures. The small change in the images when they are uncompressed is not noticed by most people.

At this point we will take a little time to delve into the details of how numbers are represented in computers because there are important practical implications. Up until now we have discussed how ADCs convert a voltage range into a binary number by assigning 0 to the bottom of the range, and the highest number that the ADC can generate (4,095 for a 12-bit ADC, 65,535 for a 16-bit ADC) to the top of the range. This *straight binary representation* of the voltage is logical but it is not the only way that the ADC can represent the voltage. One immediate problem with the straight binary representation occurs when we switch from unipolar to bipolar mode. In unipolar mode everything is fine. An input of 0 V generates a code of 0, and an input at the top of the range generates a maximal code. The host computer can readily convert a binary code into a voltage by multiplying the binary code by the ADC step size. However, this falls apart when the ADC is used in bipolar mode. Now the lowest code represents a negative voltage, and 0 V is represented by the mid-range code. The host computer has to use a different conversion system to change ADC codes into an equivalent voltage.

What we need in bipolar mode is a method for representing negative voltages. One method is the same as that used in ordinary arithmetic. One bit is used to represent the sign of the number: 0 for a positive number and 1 for a negative number. The rest of the binary number represents the magnitude of the voltage. This system, called *sign magnitude representation*, is not widely used for a couple of reasons. Firstly, it turns out that it is difficult to perform binary arithmetic on sign magnitude numbers; in particular, addition and subtraction are not the same (i.e. adding two positive numbers is performed differently from adding a positive and negative number). Secondly, there are two zeros: $+0$ and -0. To get around these problems all computers use a different method of representing binary numbers called *2's complement representation*. In this system, positive numbers are represented by the straight binary equivalent. Negative numbers are represented by the binary number that you need to add to the positive number of the same magnitude to get 0. Thus, addition and subtraction are the same process in 2's complement arithmetic. To form a negative number in 2's complement format, take the binary equivalent of the same positive number, complement each bit (change 1 into 0 and vice versa) and add 1. In 2's complement format, all positive numbers start with the binary digit 0, and all negative numbers start with 1. There is only one 0, represented by all the bits being 0. It turns out that 2's complement format is the same as straight binary representation except that all negative numbers have the first digit as 1 instead of 0.

What relevance does all this have to ADCs? Well, most ADCs use 2's complement format when in bipolar mode and straight binary when in unipolar mode. If you are writing programs that deal with ADC data on a low level it is vital that this is appreciated and the appropriate conversion algorithm is used, otherwise the results will be nonsense. High-level languages deal with the conversion automatically but this only works if the ADC gathering the data behaves in the way that the program expects and the two can communicate, which is a good reason for buying hardware and software from the same manufacturer. The change between unipolar and bipolar mode on many low-cost ADC boards is performed manually – you have to set a switch or move a jumper on the ADC board. The data acquisition software often does not know which mode the ADC board is in because there is no facility for communicating this information between the two, so it is up to the user to tell the software what mode the ADC is set to. If the software and the ADC board are not set to the same mode the data that are collected will be meaningless.

If you need to work with low-level ADC data there are further traps awaiting you. A 12-bit ADC only needs 1.5 bytes to represent a voltage but because computers prefer to deal in whole bytes the code is represented as a two-byte (16-bit) number. The problem is that there is more than one way of fitting a 12-bit number into 16 bits. You can either leave the number as it is and add four zeros onto the leading edge (most significant bits) of the number, or add four zeros to the trailing edge (least significant bits) of the number. The latter effectively multiplies the number by 16. Both systems are used by ADC manufacturers and you have to know which one applies to your specific ADC. Obviously, this can cause havoc if you write a program for one type of ADC and it is used with another.

Quite often, the reason that one has to work with low-level ADC data is in order to transfer raw data from one computer to another. For example, imagine that a piece of complex analysis software has been written for a particular computer. The computer is now outdated and will not support new ADC hardware. The software cannot be moved to a newer computer without a major reprogramming effort, so the data are collected on new computers and analysed on the old one. Other scenarios include collaborative work where experiments are performed in one laboratory and the data sent to another for analysis. In all these cases, not only do you have to ensure that the format of the raw data is correct but you also have to transfer the data between computers. Inevitably, this process is not without pitfalls. One problem that you will encounter is the way in which different computers deal with binary numbers internally. Some computers have the most significant bit first, in the same manner as ordinary decimal numbers. This format is known as *big-endian*. Other computers have the least significant byte first, which is *little-endian* format. Normally this difference is completely hidden to the user. However, problems arise when transferring data from one machine to another because computers save data in the same order as they process them. Thus a big-endian computer saves a two-byte number as high byte first, low byte second. If the file is then read by a little-endian machine it will interpret the first byte that it reads as the low byte and the second as the high byte. This error causes havoc! It is an important issue because endian differences between computers are common. In particular, PC-compatible computers are little-endian and Macintoshes are big-endian so byte-order conversion is needed when swapping data files between these machines. It is easy to forget this because both computers will read discs written by the other (on the PC you may need a utility program to do this). The file read/write routines provided by the computer's operating system do not perform byte-order conversion and you have to specifically add a conversion routine. The conversion procedure is straighforward but low level. You have to reverse the order of the bytes in each 16-bit number, swapping the high and low bytes. Such low-level number and file manipulation requires reasonable programming skills but you do not have to be a Windows guru to do it. The program can be written in a data analysis language that provides file manipulation routines such as LabView.

Transferring two-byte (16-bit) data between computers is bad enough, but transferring floating-point data is a nightmare! This is because there are several different ways of representing floating-point numbers in binary. The representation varies depending upon the precision of the number and the type of computer. For example, an extended precision number (the highest precision format for a floating-point number) occupies eight bytes in Windows, twelve bytes on older Macintoshes and sixteen bytes on Power Macs and Sun machines. Although there is a standard format for representing floating-point numbers (the IEEE format) it is not always followed, especially on older machines that were built before the new standard was written. In addition, you still have to know what precision was used when the numbers were stored. In practice, these differences make it very difficult to transfer binary-format floating-point data between different types of computers.

There is, however, another way to transfer data that is universal and extremely useful. The data are stored in text (ASCII) format. This means that the numbers are written to disc in exactly the same way as we would write them on a piece of paper. For example, the first six digits of the number π would be written as a string of digits plus a full stop (3.14159), not as a block of twelve bytes of binary data. The huge advantage of this system is that the code used to represent letters and numbers on computers (ASCII code) is standard across all machines. The actual process of converting a twelve byte binary number into a string of digits and back again is complex but it is built into all computers because this is how they display numbers on the screen. Another advantage of storing numbers in ASCII format is that they can be manipulated easily by user-friendly programs such as Microsoft's Excel. Excel will read any text file and attempt to convert the contents into numbers in a spreadsheet. The rules for storing numbers in a format that can be read by Excel and other data-handling programs (graphic and statistical packages) are simple. The numbers themselves are in ASCII format. Numbers are separated by a 'pad character' (usually a comma for PC programs and a tab for Mac programs but the 'wizard' in Excel will cope with most non-numeric characters). When Excel reads the file it places successive numbers in one row with each new number to the right of the previous one. To start a new row an 'end of line' marker is used instead of the pad character. The end-of-line marker is a carriage return (CR) character in Macs, a line feed (LF) character in UNIX and a carriage return followed by a line feed (CR/LF) in PCs. However, Excel's Wizard will cope with any of these. Data in ASCII format are also easier to send by e-mail because they are in text-only format and if necessary can be embedded into the body of a message.

DIGITAL-TO-ANALOGUE CONVERTERS

The opposite of analogue-to-digital conversion is, not surprisingly, digital-to-analogue conversion. Many data acquisition/analysis systems include one or two *digital-to-analogue converters* (DACs) on the circuit board. They allow the computer to generate signals from stored or calculated values. DACs work just like ADCs in reverse; the computer sends them a number and they produce a corresponding voltage. Like ADCs, DACs have a range that is divided up into a number of equal steps. The number of steps is described in terms of bits in the same way as ADCs so that the number of steps $= 2^{(\text{no. of bits})} - 1$. For example, a 12-bit DAC has

4,095 steps. If this DAC has a range of -5 V to $+5$ V then each step is equivalent to $10/4,095 = 0.00244$ V. Sending 0 to the DAC would produce -5 V at the output and sending 2,048 to the DAC would produce 0 V at the output.

One obvious use for a DAC is if you want to 'replay' signals that you have recorded, like a tape recorder. The process is simple:

i. set the range of the DAC to the same as that used by the ADC when the data were collected;
ii. send each sample to the DAC at the same rate as that used to sample the data.

You can see that there is considerable potential even in this relatively simple process. For example, suppose that a monitoring system records air temperature over a 24-hour period. The air temperature does not change very quickly so a slow sampling rate of one sample every 10 seconds can be used, giving 8,640 samples over a 24-hour period. If the data are sent to the DAC at 288 samples/s and the output of the DAC connected to a chart recorder running at a speed of 1 cm/s the whole 24-hour record will fit on 30 cm of chart-recorder paper, giving a very convenient overview of the whole 24-hour period.

DACs are also commonly used to generate a signal. One advantage of using them over a standard signal generator is that you can generate any waveform that you like, no matter how complex. As a simple example, suppose a 100 Hz sine wave, 2 V peak-to-peak amplitude (Figure 2.8), is needed. Twenty samples per cycle will give us a good representation (see Chapter 3 for more details). To work out what numbers should be sent to the DAC, first work out what the voltage will be at each of the 20 points. A cycle of a sine wave is 360°, so dividing this range up into 20 steps gives $360°/20 = 18°$/step. The peak-to-peak amplitude is 2 V, so the maximum value (positive or negative) is 1 V. Combining all these ideas, the voltage at each point is given by

$$V_i = 1.0 \times \sin(i/20 \times 360)$$

where $V_i =$ the voltage at point i. The actual numbers are shown in Table 2.1.

These numbers now have to be scaled to the correct digital value. The range of the DAC is -5 to $+5$ V and it is a 12-bit DAC, so a digital value of 0 gives -5 V and a digital value of 4,095 gives $+5$ V. The number that we need to send to the DAC to get a certain voltage is given by

$$\text{digital number} = (\text{voltage} + 5)/10 \times 4,095$$

These numbers are shown in Table 2.2.

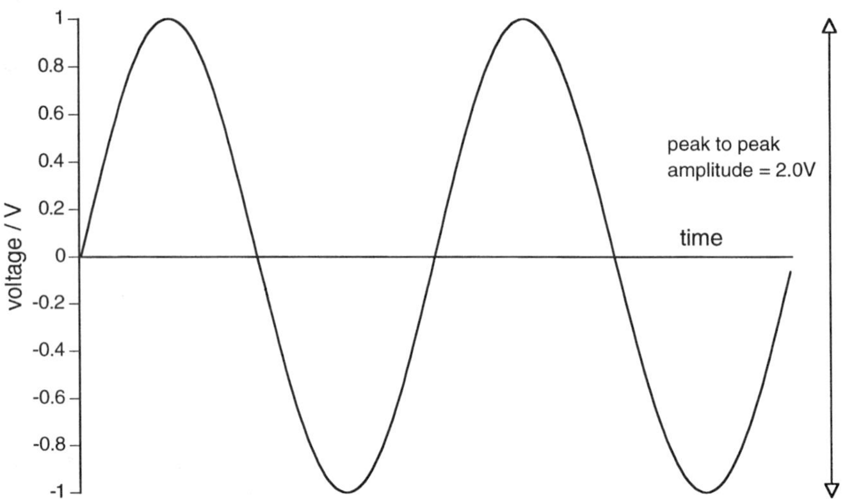

Figure 2.8 A sine wave with a peak-to-peak amplitude of 2.0 V.

TABLE 2.1 ONE CYCLE OF A SINE WAVE WITH 2.0 V PEAK-TO-PEAK AMPLITUDE, DIVIDED INTO 20 SAMPLES

No. of sample (i)	Voltage (V)	No. of sample (i)	Voltage (V)
1	0.31	11	-0.31
2	0.59	12	-0.59
3	0.81	13	-0.81
4	0.95	14	-0.95
5	1.00	15	-1.00
6	0.95	16	-0.95
7	0.81	17	-0.81
8	0.59	18	-0.59
9	0.31	19	-0.31
10	0.00	20	0.00

How quickly should the values be sent to the DAC? The frequency of the sine wave is 100 Hz, and in each cycle there are 20 points. The samples need to be sent to the DAC at $20 \times 100 = 2{,}000$ samples/s.

The resultant waveform is shown in Figure 2.9. Note that it is not quite a pure sine wave, but has a 'staircase' appearance, which is due to the way the synthesized sine wave was produced. As soon as each new number is sent to the DAC the output changes to its new value and remains there until the next sample is sent

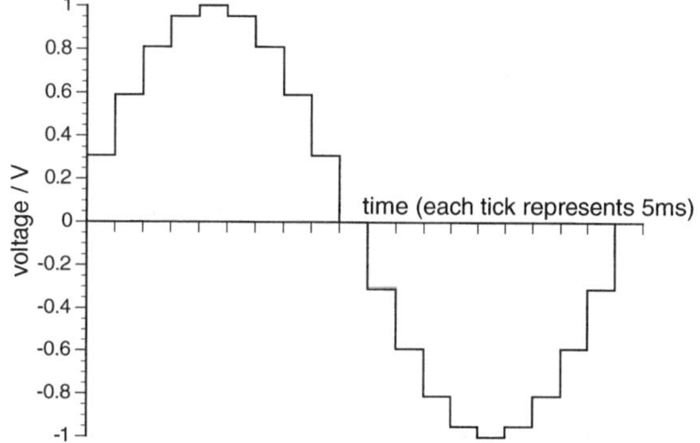

Figure 2.9 The 'staircase' appearance of a sine wave from a DAC. The 'steps' can be smoothed out with a low-pass filter to give a smooth sine wave.

TABLE 2.2 **THE 20 SAMPLES IN TABLE 2.1 CONVERTED FROM VOLTAGE VALUES TO NUMBERS TO BE SENT TO A 12-BIT DAC WITH A RANGE OF -5V TO $+5$V**

No. of sample (i)	Digital number	No. of sample (i)	Digital number
1	2175	11	1921
2	2289	12	1807
3	2379	13	1717
4	2438	14	1658
5	2458	15	1638
6	2438	16	1658
7	2379	17	1717
8	2289	18	1807
9	2175	19	1921
10	2048	20	2048

5 ms later. A low-pass filter (Chapter 3) is needed to 'smooth out' the staircase effect and give a true sine wave.

DIGITAL INPUT AND OUTPUT

Data acquisition/analysis systems often include the facility to deal directly with digital signals. This is usually described as digital input and output, abbreviated to *digital i/o*. Digital i/o circuits move information in the real world into the computer and get information from the computer into the outside world. In this sense they are like ADCs and DACs, 'translating' between the real world and the computer. But the task of digital i/o circuits is much easier than the analogue ones because both worlds are talking the same language: digital. All the digital i/o circuits have to do is clean things up a bit. The circuitry to make a digital input is so similar to a digital output that the two functions are combined into one, hence the term digital i/o. A digital i/o circuit provides a set of connections which can either be inputs or outputs: you choose which you want them to be. The connections are set to the appropriate mode (input or output) for you during an *initialization* process. During initialization you tell the computer which connections you want to be digital inputs and which you want to be outputs. The computer then sends the digital i/o circuits instructions which set the connections to their correct mode of operation. This initialization process is performed at the start of an experiment: you usually do not want to change the setup in the middle of an experiment. Like ADCs, setting up digital i/o connections can be tricky, so the manufacturers of data acquisition systems provide driver programs to do this.

Digital i/o circuits are described by the number of connections that they provide. Each connection is called a *line*, and they are nearly always in multiples of eight. 'Sixteen lines of digital i/o' is a common way of describing that a data acquisition system has sixteen digital connections available, each one of which can be either a digital input or a digital output. A collection of i/o lines is called a *port*.

How are digital i/o connections used? Computers deal only with numbers. These numbers are represented inside the computer itself by different voltage levels. Computers work on binary numbers, so only two different voltages are needed, because everything that computers do is described inside them in terms of ones and zeros. The system used in all computers is really simple – a high voltage represents a 1 and a low voltage represents a 0. The definition of 'high' and 'low' varies depending on the technology that is used to make the computer but a common technology called CMOS represents a 0 by a voltage less than 1.5 V and a 1 by a voltage greater than 3.5 V (when running off a 5 V supply).[5] In between these two levels is 'undefined', and the voltage should not be in this range because the computer does not know whether to call it a 1 or a 0. In practice, the low voltage used is about 0.05 V and the high voltage is 4.95 V to give a 'safety margin'. Another common technology is TTL, which represents a 0 by a voltage less than 0.8 V and a 1 by a voltage greater than 2.0 V.[6] The digital i/o ports of data acquisition systems are described by statements such as '16 lines of CMOS compatible digital i/o'. As long as you follow the rules for CMOS technology, which means that a voltage less than 1.5 V represents 0 and a voltage greater than 3.5 V represents 1, you and the computer will understand each other perfectly. Digital i/o is like a window straight into the heart of the computer. Instead of having to

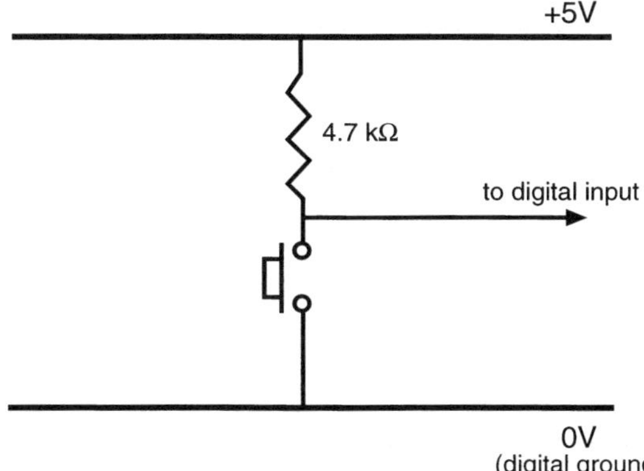

Figure 2.10 A simple way of wiring up a digital input so that the computer records a '1' when the switch is open and a '0' when the switch is closed.

use an ADC to represent a voltage as a number, you can talk directly to the computer in its own language. There are only two numbers (1 and 0) in the digital system used by computers, so digital i/o works best for controlling or recording things that have two states: on/off, yes/no, start/stop, awake/asleep, fast/slow, and so forth.

An example of the use of a digital input is to record when a button is pressed: this could be used to mark an event, for example, the point in the experiment when a drug was given. Pushing the button needs to be converted into a change in voltage that the computer can record. A simple way to do this is shown in Figure 2.10. A separate 5 V power supply or battery is not needed because every digital i/o board will have a connection to the computer's own 5 V supply. When the switch is open the input is at +5 V and the computer will record a 1. When the switch is closed the input is at 0 V and the computer will record a 0.

For many applications the output needs to be boosted with a transistor because the power available from the board is rather limited (e.g. CMOS technology can supply a maximum of 0.5 mA).

Digital output is just as easy. Again, it is most suitable for on/off applications. Figure 2.11 shows a very simple example of using a digital i/o line to turn a light on and off: this might signal that a certain stage has been reached in the experiment. A light emitting diode (LED) makes a cheap and reliable lamp – ideal for this purpose. In Figure 2.11a the light comes on when the digital output is set to 1, and in Figure 2.11b the light comes on when the output is 0. LEDs need 5–10 mA of current flowing through them to be bright enough to see and it is important to make sure that the i/o circuit can supply it. The circuits in Figure 2.11 use about 7 mA of current, set by the 470 Ω resistors. The manufacturers will specify the maximum current that the board can supply. Often, two figures are given: the

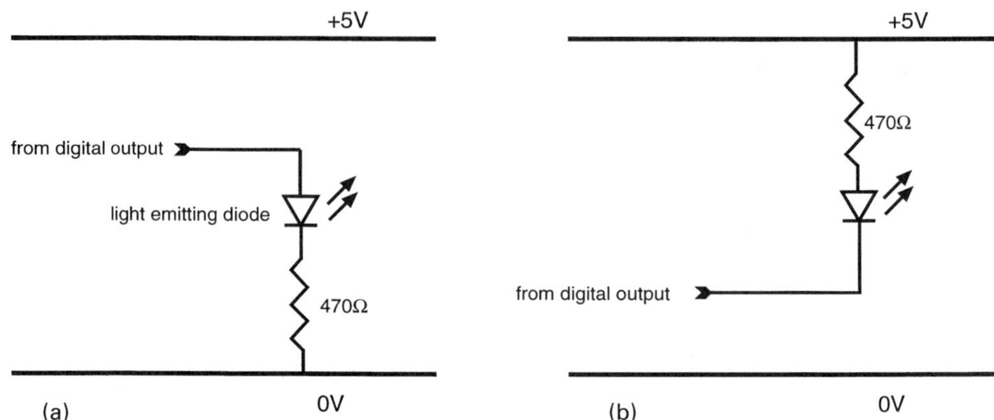

Figure 2.11 Using a digital output to directly turn a light-emitting diode on and off. In (a) the light comes on when the digital output is set to 1, and in (b) the light comes on when the output is 0. The digital output must be able to supply at least 7 mA of current.

source current and the *sink* current. The latter is usually greater. The output sources current when the load is connected between the output and ground, as in Figure 2.11a. Conversely, the output sinks current when the load is connected between the output and the +5 V supply (Figure 2.11b). If you need to control more powerful devices with digital outputs you will need something to boost the current. Figure 2.12 shows a 'minimalist' solution to this problem that will source up to 200 mA with one extra component. The 2N7000 MOSFET transistor (available from Digi-Key and Allied Electronics) performs this task by acting as a voltage-controlled switch. The two terminals of the switch are the drain and the source, and the gate controls the switch. If the voltage on the gate is 0 V then the switch is open. There is effectively no connection between the drain and the source and so the LED is off. If the voltage on the gate is about +5 V the switch is closed; there is a good connection between the drain and the source and the LED lights up. The 180 Ω resistor sets the current through the LED at about 20 mA, which gives a bright light. The useful feature of the 2N7000 transistor is that the gate takes virtually no current to open and close the switch. Note that this circuit is 'inverted' compared to Figure 2.11b; the LED is on when the digital output is set to 1. If you have trouble finding the 2N7000 transistor you can use a general-purpose NPN transistor such as the PN2222, also available from Digi-Key (Figure 2.13).[7] An extra 10 kΩ resistor is needed because the PN2222 is a bipolar transistor, not a MOSFET like the 2N7000. Wiring up individual transistors to digital outputs becomes tedious when more than a few lines are needed and there are many solutions to the problem, either using multiple transistors in one

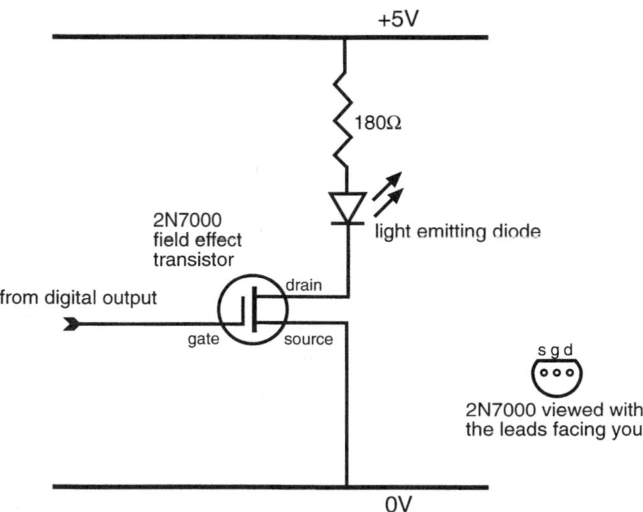

Figure 2.12 A 2N7000 MOSFET transistor is used to boost the output of the i/o circuit so that a higher-power LED can be used. In this circuit the LED takes 20 mA from the power supply but almost nothing from the digital output.

Figure 2.13 The same circuit as Figure 2.12 but using a PN2222 bipolar transistor. An extra $10\,k\Omega$ resistor is needed to set the transistor's base current. This circuit takes about 0.4 mA from the digital output.

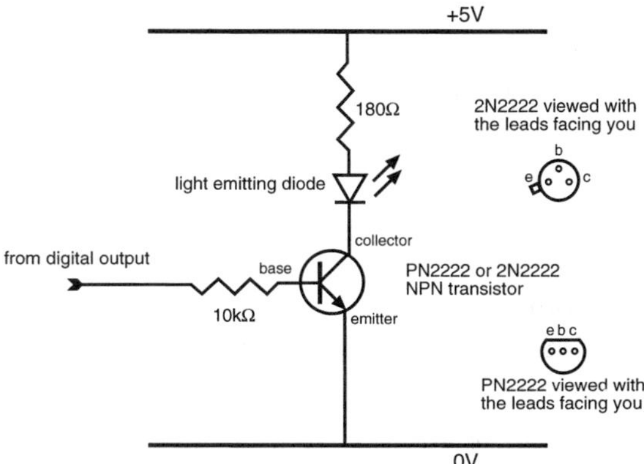

package or complete 'buffer' integrated circuits that need no additional resistors. If you really need to drive something large there are high power MOSFETs available such as the IRL530N (available from Digi-Key) that will sink several amperes. Unfortunately, the details of the design and construction of such circuits are beyond the scope of this book. A good place to look for ideas is in Horowitz and Hill's book (pp. 582–8) which gives several circuits that use digital outputs to drive lights, relays, and so forth. There is also a wide range of digital output boards available from ADC manufacturers that have these circuits built in.

A word of caution with digital i/o. It is possible to damage the i/o circuitry, even if the voltages used are within the correct range for the board (this range is nearly always 0 to +5 V). This only happens if the digital i/o line is acting as an output, and the most common reason is if, by mistake, it is treated as an input. Look at Figure 2.10 again. Suppose that instead of making the i/o connection an input it is made an output, and set to give the value 1. This means that the i/o circuit sets the line to 4.95 V. If you now press the pushbutton, the i/o line is connected directly to 0 V, short-circuiting it. This short-circuit may permanently damage the data acquisition board. This type of damage only occurs when the digital i/o line is set up as an output, which is why all i/o lines are automatically set up as inputs when the power is first turned on to the data acquisition system. It is relatively easy to prevent such mishaps by careful design of the circuits attached to the i/o lines, at the minor expense of a few more components. Besides the short-circuit protection it is advisable to build in over-voltage and static discharge protection at the same time. A simple way to provide all of this is shown in Figure 2.14. The 330 Ω resistor limits the current to a safe value in the case of a short-circuit. The diodes provide protection from static and moderate excess voltages. Normally the diodes do nothing, but if the voltage applied to the 330 Ω resistor goes above 5.6 V or below −0.6 V they start to conduct and shunt the excess current into the power supply, which can cope much better with small overloads than the digital i/o. Almost any 1 A, 50 V diode (e.g. the 1N4001, available from Digi-Key) will work.

Finally, digital i/o lines can be *isolated*. This means that there is no direct electrical connection between the ADC board and the external devices. Instead, the connection is made optically using an opto-isolator. The lack of a direct connection protects one device if the other one fails or blows up, which saves your computer or apparatus from serious damage.

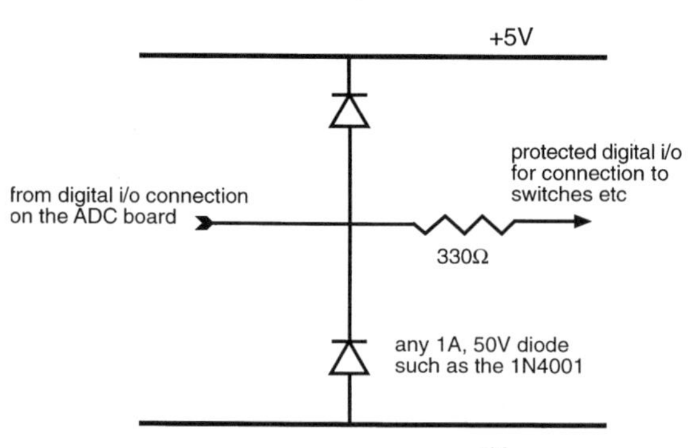

Figure 2.14 A simple protection circuit for digital i/o ports. It provides protection from short-circuits, static discharge and moderate voltage overloads.

COUNTER/TIMER INPUTS

A variation on the basic digital i/o port is the *counter/timer*. Most data acquisition systems with a digital port have one or two counter/timers. These are digital inputs and work over the same range of voltages as standard digital inputs. Like basic digital i/o lines, the counter/timers can be configured either as inputs or outputs. What is different about counter/timer inputs compared to digital i/o is that the i/o hardware does some processing itself, rather than relying on the host computer for everything. Not only does this free up the computer for other tasks but the i/o circuitry can perform its simple tasks much faster than the host computer. Setting up the counter/timer lines can be somewhat confusing because of the large number of options available but this flexibility does help avoid having to add on extra interface circuitry.

In counter mode, the line is configured as an input. The counter counts pulses applied to it: a single pulse is counted when the digital input goes from low to high and back to low again (or from high to low to high – one of the many options available!). The number of pulses is then read by the host computer, which can also clear the counter (set the count to 0). Typically, a stream of pulses from an external device is applied to the counter input. At regular intervals the host computer reads the counter and clears it. The computer now knows how many pulses there were in the time period and can thus work out the pulse frequency. For example, one application is to count the speed of rotation of a device. The device could be an anemometer, for example, whose rotation rate is proportional to the speed of the air flowing over it. Attached to the anemometer is an optical encoder (see Chapter 4) that sends out a pulse every time the anemometer moves by $3.6°$ (100 pulses per revolution). It is now easy, knowing the calibration factor of the anemometer, for the computer to calculate the air speed.

The counters in the i/o circuitry are digital and binary. They are specified by the number of bits in each counter. For example, the 'E' series A/D boards from National Instruments have two 24-bit counters that will count up or down (24-bit means that the counters count from 0 to $2^{24} - 1$, i.e. 16,777,215, after which they start over again at 0). There is no overflow warning and it is up to you to watch for it. This may not be a problem if the number of pulses cannot possibly approach 16,777,215 in the time that the program will run for and all you have to do is clear the timer when the program starts up. Although it may seem sensible to clear the counter each time it is read, this is not usually done because you may miss some pulses in the time it takes the software to send the 'clear' command to the counter. Instead, the counter keeps going and each time the counter is read the previous count is subtracted from the current one. If the counter overflows it is easy to detect because the new value will be lower than the old one. To get the true increase in pulse count after an overflow, add on the maximum count plus one (i.e. 2^{24} for a 24-bit counter). To illustrate this, suppose the new count that you have just read is n_{new} and the old one is n_{old}. The number of pulses that occurred in the interval between the two reads is $n_{\text{new}} - n_{\text{old}}$. Normally this will be a positive number. If it is negative, add on the maximum count plus one, so for a 24-bit counter the number of pulses after an overflow is $n_{\text{new}} - n_{\text{old}} + 2^{24}$.

In timer mode the counter/timer acts like a digital stopwatch. The counter is connected internally to a circuit that generates a stream of pulses of known frequency. This circuit is often referred to as a 'clock'. In the 'E' series boards the pulse rate (the 'clock rate') is either 20 MHz or 100 kHz. Whether the counter counts the pulses or ignores them is controlled by the state of the timer line. For example, the counter may only count when the timer line is high and not low. In this way the timer can measure the width of a pulse applied to the timer line. When the line is low the timer is cleared. When the line goes high the counter counts clock pulses until the line goes low again, when it stops. The value of the counter is the number of clock pulses that occurred whilst the line was high. Since the clock rate is known (20 MHz or 100 kHz for the 'E' series boards) the width of the external pulse applied to the timer line is easy to calculate. The clock rate that you use depends upon the expected width of the external pulse. The resolution to which you can measure the width of the pulse is ± 1 clock pulse, which is 50 ns for a 20 MHz clock and 10 µs for a 100 kHz clock. On the other hand, you do not want the counter to overflow and so the maximum external pulse width is 839 ms for a 20 MHz clock and 168 s for a 100 kHz clock. For the best overall resolution you use the counter/timer in count mode for high-frequency pulses and timer mode for low-frequency pulses.

Timer/counter lines can also be configured as outputs. In this mode you can use them to generate pulses of accurately defined width. For example, you can set up the counter/timer so that the output goes high for a certain number of clock pulses and then goes low again. If the number of clock pulses is set to 2,000,000 and the clock rate is 20 MHz the output will go high for 100 ms, accurate to ± 50 ns. Often this is done by setting the counter to 2,000,000 and then counting down; when the counter reaches 0 it resets the output line to low.

The number of options available with counter/timer lines makes them a little daunting to use at first, but they can be extremely useful and their high speed extends the capabilities of the data acquisition system. One application that I worked on recently involved counting the rate that motor neurons fired in response to a stimulus. Each time a motor neuron fired, it generated a small EMG spike. The EMG spikes were amplified and fed to a pulse height discriminator that generated a digital pulse each time the EMG exceeded a pre-set level. The digital pulses were fed to the counter input of the data acquisition system and the counter was read by the computer once every second. By using the counter/timer input the computer only had to make one reading each second, rather than counting the thousands of spikes that occurred during this period.

chapter three

Important Concepts

This chapter introduces some important ideas that are fundamental to digital signal processing. If they are new to you take your time, the concepts are not easy to grasp and are best taken in small doses. An understanding of how digital systems work is not essential in order to use them but some knowledge is useful for solving problems and obtaining optimum results.

SAMPLING RATE

The concept of sampling rate was mentioned earlier. This important subject is now discussed in more detail.

What happens when a signal is sampled?[1] A series of 'snapshots' of the signal is taken at regular intervals. The signal itself is continuous and represents some variable such as flow or pressure. The sampled signal is a string of regularly spaced 'snippets' of the continuous signal. This is very similar to the way a movie camera works: it starts with a continuous picture, that is, the real world, and takes a stream of still photographs at regular intervals. Since all digital data acquisition systems work with sampled signals rather than the continuous ones, an important question is 'how well can a sampled signal represent a continuous signal?' The answer is rather surprising.

Figure 3.1 shows a continuous signal coming from a transducer. Imagine that this is connected to a switch that is closed for a brief period at regular intervals as shown in Figure 3.2. When the switch is closed the signal passes through and when it is open nothing gets through, rather like a movie camera shutter which lets light through when it is open and nothing when it is closed. The result is the sampled signal shown in Figure 3.3a. This process can be analysed mathematically to see in what ways the sampled signal resembles the original continuous signal. It turns out that, *provided the signal is sampled above a certain rate, all the*

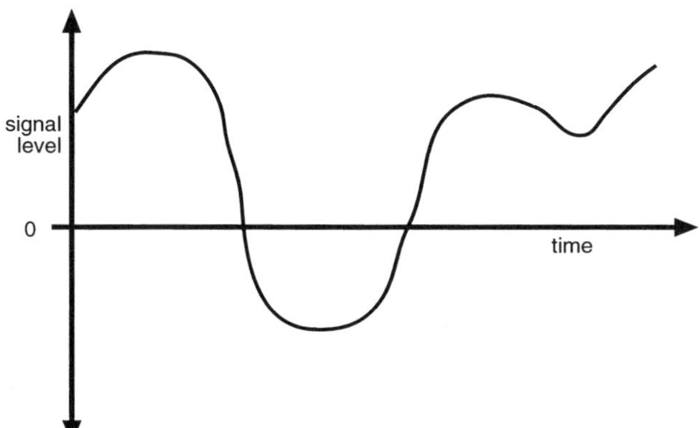

Figure 3.1. An electrical signal coming from a transducer. A 'signal' is a voltage that varies with time.

Figure 3.2 A continuous signal is sampled by an electronic switch that closes for a brief period of time at regular intervals. The number of times per second that the switch closes is called the sampling rate.

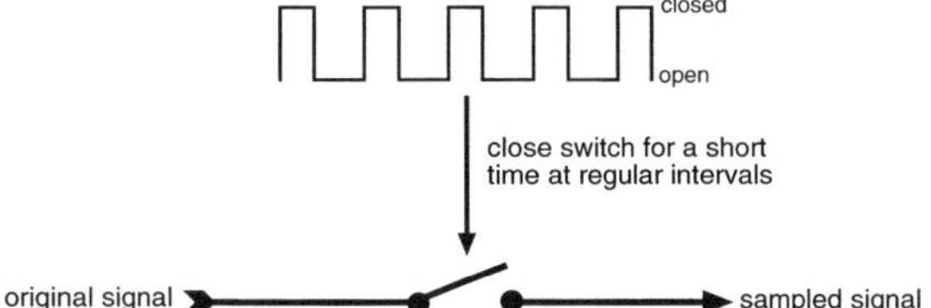

information present in the original signal is preserved in the sampled signal. This is quite amazing – by only taking a quick peek every now and then at what is going on we know the whole picture. The analogy with the movie camera is useful again: a movie is a stream of snapshots (24 per second) of the real world, yet what we visualize certainly seems close to the original scene.

One very important point that leads to a lot of confusion is that the sampled signal is *not* the same as the original (Figure 3.3a). All the information in the original signal is contained in the sampled version, but there are some extra bits and pieces as well. These can be removed, if necessary, to give the original signal but usually once a signal has been sampled all subsequent procedures are performed on the sampled version. A data acquisition system converts a real-world signal into a sampled and digitized version, which consists of a string of numbers spaced at regular intervals in time. This digitized version of the signal is frequently depicted by drawing a graph of the numbers against time (Figure 3.3b) with the individual samples joined by lines. This is invariably how sampled signals are shown on computer monitors. Although this is a very useful way of representing the signal, it is just that: a representation of the sampled signal. It is

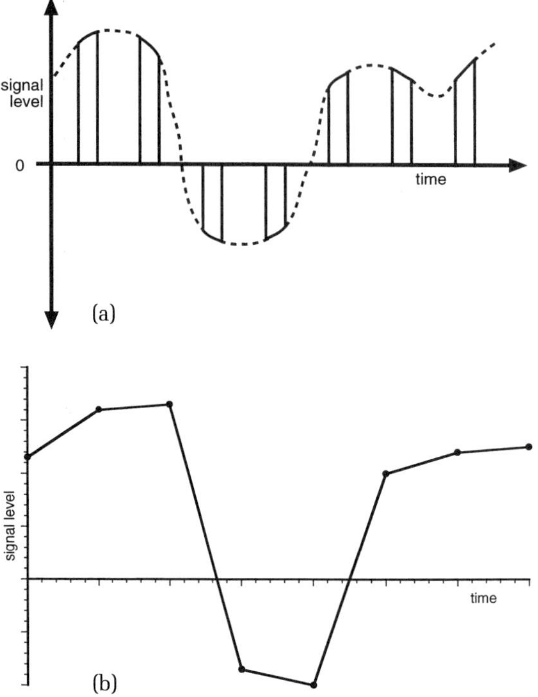

Figure 3.3 (a) The result of sampling the signal in Figure 3.1. When the switch is closed the voltage is the same as the continuous signal and when it is open the voltage is zero. The solid line is the sampled signal and the dashed line is the original continuous signal. (b) The way that sampled and digitized signals are frequently depicted. The solid circles are the samples themselves: the lines joining them are added and are artefactual.

not the same as the sampled signal (which has nothing in between samples) and it is definitely not the same as the original signal. Near the end of the chapter are some examples of how the differences between original and sampled signals can lead to problems if not appreciated.

To see how sampling works and to investigate the effects of varying sampling rate, the concept of *spectral analysis* now needs to be introduced. Spectral analysis is covered in detail in Chapter 5 and only a few of the ideas are used here. All signals, no matter how complex, can be made up by adding together a number of pure tones or sine waves. Figure 3.4 shows how a fairly complex-looking signal is made up from the sum of three sine waves. Each sine wave has three characteristics: its *amplitude* (the height of the wave), *frequency* (how many waves there are per second) and *phase angle* (where the wave starts relative to others). The three sine waves in Figure 3.4 have amplitudes (top to bottom) of 1, 1/3, and 1/4 arbitrary units. Their frequencies are 1, 3, and 5 units: three cycles of the second wave and five of the third fit in the space of one cycle of the first wave. The phase relationships are 0°, 90° and 0°: the first and third waves start at the mid-point of a cycle, whereas the second wave starts at the peak of a cycle. The *frequency spectrum* of a signal is just a graph showing the amplitude and phase of the sine waves

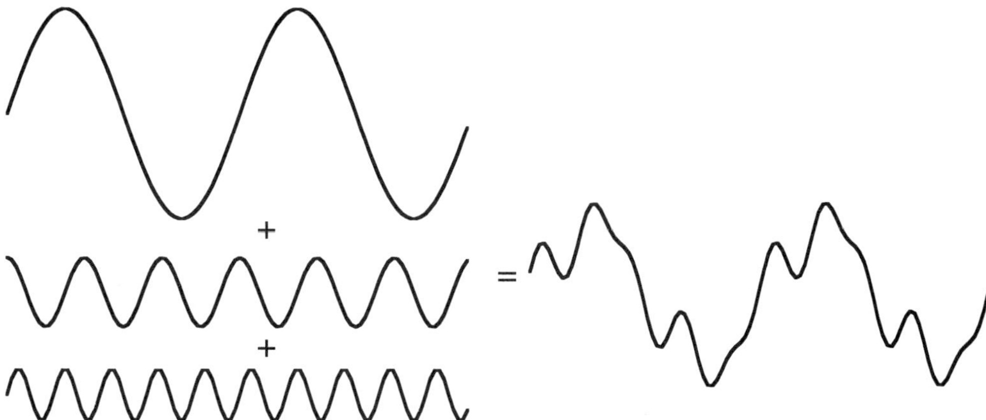

Figure 3.4 The signal on the right is made up from the sum of the three sine waves on the left. The amplitude and phase angles of the three sine waves are, from top to bottom: Amplitude 1, phase angle 0°; Amplitude 1/3, phase angle +90°; Amplitude 1/4, phase angle 0°.

Figure 3.5 The amplitude spectrum of a signal is a graph of the amplitude of the sine waves that make up the signal plotted against their frequency. This is like taking a chord played on the piano and plotting a graph of the loudness of each note making up the chord against its position on the keyboard.

that make up the signal plotted against their frequency. Quite often, only the amplitudes are of interest and the phase angles are ignored. A good analogy is with music. Suppose that a piece of music is being played on a piano and at a certain point in time a graph of which keys are being pressed is plotted (Figure 3.5). The keys are arranged along the horizontal axis with the lowest-frequency notes on the left and the highest-frequency notes on the right, just as they are on the keyboard. In the example, four keys (C, E, G, and C′) have been pressed. The height of each line (vertical axis) represents how loudly each note is

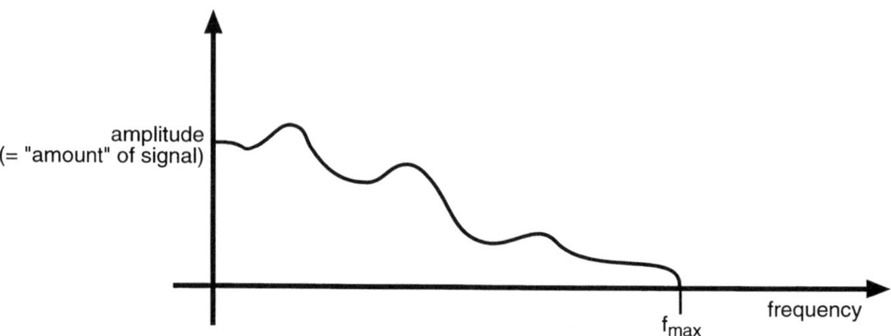

Figure 3.6 The frequency spectrum (amplitude only) of the continuous signal shown in Figure 3.1. Here f_{max} is the highest-frequency component present in the signal.

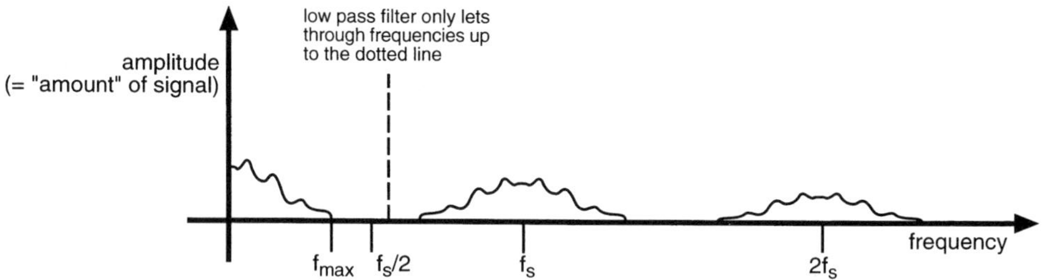

Figure 3.7 The frequency spectrum of the sampled signal shown in Figure 3.3. The sampled spectrum consists of the spectrum of the continuous signal repeated at the sampling frequency f_{s}. f_{max} is less than $f_{\mathrm{s}}/2$ so there is no aliasing. The original signal can be recovered intact by passing the sampled signal through a low-pass filter.

sounded, that is, its amplitude. This graph would be similar to a frequency spectrum of the sound produced when the four keys are pressed at the same time.

Figure 3.6 shows the frequency spectrum (amplitude only) of the unsampled signal in Figure 3.1. In general, real-world signals contain many frequencies and so their spectra are usually plotted as smooth line graphs like Figure 3.6, rather than bar graphs as in Figure 3.5. The signal contains a variety of frequencies up to a certain point and nothing beyond it – this maximum frequency will be called f_{max}. The spectrum of the sampled signal from Figure 3.3 (i.e. the sampled version of Figure 3.1) is shown in Figure 3.7 – the scaling of the axes has changed a bit to fit it on the page. Note that the first part of the spectrum is the same as the spectrum of the original signal – all the information in the original signal has been preserved. There are also some extra bits at higher frequencies. The sampling frequency, f_{s}, is the rate at which the switch was closed to take the samples. The extra bits consist of the spectrum of the original data and its mirror image repeated at multiples of the sampling frequency. These additional components have been added by the sampling process. They are not very useful because they do not contain any extra information, but as long as they do not interfere with the first part of the spectrum they do not do any harm. All the information needed to construct the original signal is contained in the first part of the spectrum and can easily be recovered by removing the extra data. To do this, a device is needed that removes everything except the first part of the spectrum. This is quite easy to do because the unwanted data are at higher frequencies (they are further to the right in Figure 3.7), so something that removes high-frequency signals but lets the low-frequency ones through can be used. Such a device is a *low-pass filter*. As long as the filter lets through frequencies up to f_{max} and blocks the extra ones added by the sampling process the original signal can be obtained intact. There is no approximation involved with sampling: the original signal can be obtained exactly as it was sampled. In practice, there is some distortion due to imperfections in the filters, for example, but the distortion can be made as small as required.

Note also in Figure 3.7 that the signals decrease in amplitude as the frequency increases. This is an effect called 'aperture distortion' and it is linked to the width of the samples, that is, the time that the switch is closed for. Aperture distortion decreases as the sample width decreases and aperture distortion is not usually that important because, in general, A/D converters have very short sample widths.

Consider Figure 3.7 once more. What happens as the sampling rate f_s is lowered? The original signal spectrum is still there but the gap between it and the spectra added by sampling decreases. This imposes tighter restrictions on the low pass filter needed to get back the original signal. The crunch comes when the spectra overlap (Figure 3.8). Now the spectrum of the original signal is distorted and the signal cannot be retrieved. This distortion is called *aliasing* and is extremely important, because once a signal is aliased it is irrevocably altered and no amount of manipulation will restore the original signal. The critical point where aliasing just starts is when the sampling frequency is twice the maximum frequency present in the signal, that is $f_s = 2 \times f_{max}$. The frequency $f_s/2$ is often called the Nyquist frequency, after the engineer Harry Nyquist (1928).

Avoiding aliasing is the reason behind one of the most important rules of data acquisition: *the sampling frequency must be at least twice the highest frequency component present in the signal to be sampled.* If this is not observed the signal will be distorted and cannot be restored to its original form. This presents a problem, because usually the maximum frequency in a signal is not known so the sampling rate f_s cannot be determined. There are a couple of ways of tackling the problem that are usually used.

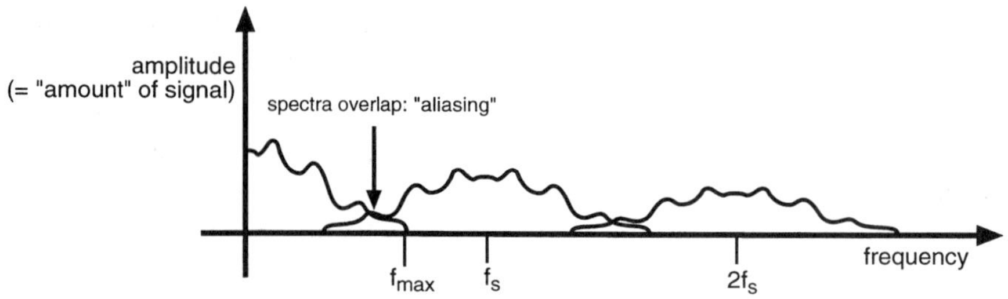

Figure 3.8 If $f_s/2$ is less than f_{max} the signal is aliased and the original signal cannot be recovered.

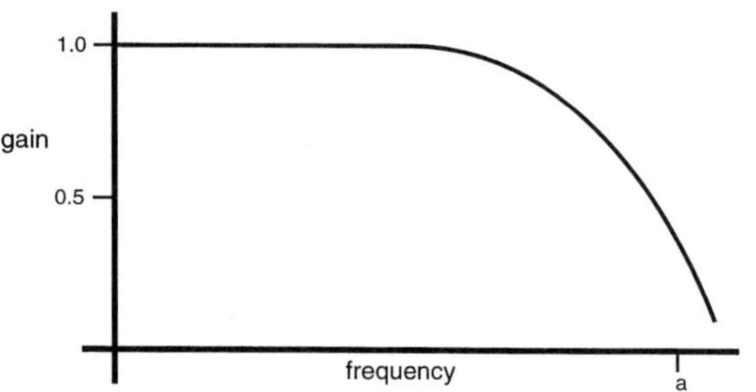

Figure 3.9 The frequency response of a low-pass filter. The gain is 1 up to some point and then it decreases steadily.

Low-pass filters are devices that do not interfere with low-frequency signals but block the high-frequency ones. If you have a hi-fi system with tone controls, put on some music and listen for a few seconds. Now turn the treble control all the way down. This makes the tone controls act like a low-pass filter and you can hear the high frequency sounds becoming more muted – the sound becomes more mellow. Figure 3.9 shows what a low-pass filter does in a more objective manner. The vertical axis is the gain, that is, the amount by which the filter alters the size of the signal going through it. The horizontal axis is the frequency. A low-pass filter has a gain of one up to a certain point, so signals in this frequency range pass through unchanged. Above this frequency the gain is less than 1 so the signal's amplitude is reduced (the signal is *attenuated*) and high frequencies are almost completely blocked by the filter. To calculate how much a particular frequency component of a signal is altered by the filter, multiply its amplitude by the gain. A gain of 1 means that the frequency component is not changed at all. A perfect low-pass filter would have a gain of 1 for signals having frequencies that are meant to pass through the filter and a gain of 0 for frequencies that are meant to be blocked. In the real world, the gain of low-pass filters never falls quite to 0 but to some small number such as 1/100,000. It is often easier to use the term *attenuation* rather than gain. Attenuation $= 1/\text{gain}$.

In Figure 3.9 there is a frequency f_a above which the gain of the filter is very small. This means that if a signal is passed through the filter the latter will block virtually all components present in the signal that have frequencies greater than f_a (this is a real-life filter so the high frequencies are not completely blocked, just reduced to an insignificant level). If a signal is passed through the low-pass filter and then sampled at a frequency at least twice as high as f_a then aliasing will not be a problem because all the frequencies present to any significant degree are below half the sampling frequency. Low-pass filters used in this way to remove any high-frequency components present in a signal before sampling are called *anti-aliasing filters*.

One problem with using anti-aliasing filters is that it is important that they do not remove useful bits of our original signal, so the frequency range of the signal needs to be known in advance in order not to lose any of it. Quite often, the frequency range of the signal can be estimated. Another simple solution is to pass the signal through a low-pass filter and see if it changes significantly. If it does not then no information has been lost.

Anti-aliasing filters are important and widely used and will be discussed in some more detail now. The problem of how to find the frequency range of a signal will be returned to later in the chapter. Figure 3.10 shows a plot of gain against frequency (the *frequency response*) of an ideal anti-aliasing filter and a real-world filter. The ideal filter lets through unchanged all frequencies below a certain point (i.e. its gain below this point is 1) and totally blocks frequencies above it (i.e. its gain in this region is 0). This ideal frequency response is often called a 'brick wall' response. In reality things are never this good and the gain of a real-world filter decreases more gently with increasing frequency. At low frequencies the gain of a real-world filter is often very close to the ideal value of 1. As the frequency is increased the gain stays close to 1 and then starts to decrease. The frequency at which the gain starts to decrease by a reasonable amount is called the *corner*

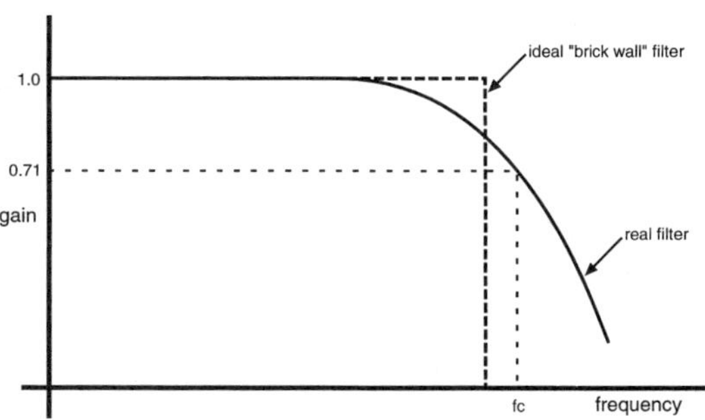

Figure 3.10 An ideal low-pass filter has a gain of 1 up to some point and a gain of 0 beyond this. The gain of real-world filters decreases more slowly. Here f_c is the corner frequency of the filter, the point at which the gain of the filter has fallen to some specified value (0.71 in this case).

frequency (also called the *cut-off frequency*). What constitutes a 'reasonable amount' varies with the design of the filter. For many types of filter the gain at the corner frequency is 0.71 $(1/\sqrt{2})$, and this is shown in Figure 3.10, where f_c is the corner frequency. The frequency range from zero up to the corner frequency is the *passband*. The gain of a low-pass filter should ideally be 1 in the passband, and how well the filter lives up to this is a measure of the *passband flatness*. The real-world filter shown in Figure 3.10 does not have a very flat passband because the gain falls to 0.71 at the high-frequency end. If it is important that the passband is flat (often it does not matter that much) then there are certain types of low-pass filters designed to have particularly flat passbands. With these types of filters the passband is usually specified to be flat within a certain tolerance, for example, ±5%, which means that the gain in this region will not go outside the range 0.95–1.05. The corner frequency will be the frequency at which the gain goes outside this range, so the gain at the corner frequency is 0.95 rather than 0.71. The degree of deviation from a flat response is commonly called ripple, so the filter can also be described as having less than 5% passband ripple.

A very important property of the filter is the rate at which the amplitude decreases above the corner frequency. The steepness of the graph in this region is called the *roll-off* rate. A filter with a steep roll-off rate is closer to the ideal 'brick wall' filter, but in general costs more. To obtain a feel for roll-off rates, imagine a point on the gain–frequency graph where the frequency is twice the corner frequency. A filter with a poor roll-off rate might have a gain at this frequency of 0.25 whereas a filter with a steep roll-off could have a gain of less than 0.004. Decent anti-aliasing filters all have steep roll-off rates.

At very high frequencies one of two things can happen to the frequency response. The gain can just continue to fall, like filter (a) in Figure 3.11, or it can level off at some low value like filter (b) in Figure 3.11, depending on the design of the filter. If the filter has a frequency response like filter (b) then there are a couple of extra terms used to describe its frequency response. The region where the gain flattens out at some low value is the *stopband*, and the amount by which the gain of the filter is reduced in this region compared to the passband is the *stopband attenuation*. Between the passband and the stopband is the *transition band*. Filter (a) does not really have a stopband, the gain just keeps on falling as the frequency increases.

The graph in Figure 3.11 is not of gain against frequency but $\log_{10}$(gain) against $\log_{10}$(frequency). Log–log graphs are very often used to describe the frequency response of filters for several reasons. Firstly, they give a nice compact format to the graph so that it is easy to understand. The gain can change by several orders of magnitude and still be easy to plot. Secondly, the gain–frequency graph of

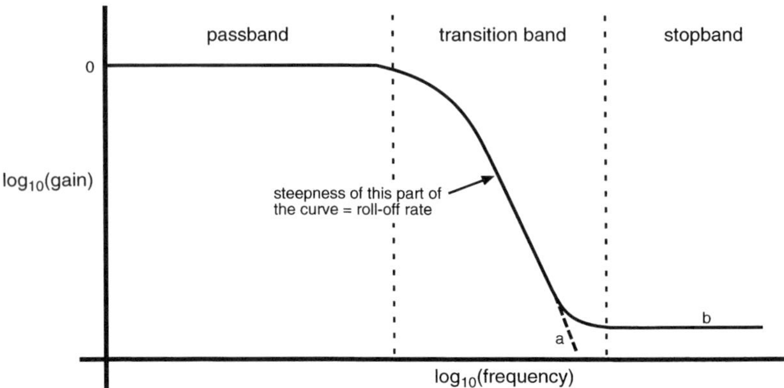

Figure 3.11 Log–log plots of the frequency response of two low-pass filters. Filter (a) has a gain of 1 (so the logarithm of the gain is 0) at low frequencies, and decreases steadily with increasing frequency. Filter (b) has a gain that eventually settles out at a small, constant value at high frequencies. The frequency response is divided into three regions: The *passband* is the region from 0 Hz to f_c. The *stopband* is the region beyond f_c in which the gain is small and constant. The *transition band* is the region in between the passband and the stopband.

several common types of filters becomes a straight line on a log–log graph above the corner frequency and so it is very easy to compare roll-off rates between different filters, which we often want to do. Thirdly, filter responses are very frequently described using a logarithmic unit, the *decibel*, and it is easy to read off values in decibels from a log–log graph.

DECIBELS

Decibels are very commonly used in electrical engineering. The decibel is not a unit in the conventional sense, that is, it is not a fundamental unit of measurement like the metre or the joule. It is a *ratio*, and thus it describes how much bigger or smaller one thing is compared to another. A lot of the confusion with decibels arises because very often the two things that are being compared are not explicitly stated: one of them is understood.

The decibel is a *logarithmic ratio of powers*. If the amount of power in two signals A and B is measured then the ratio of their powers in decibels is given by

$$\text{ratio of power (decibels)} = 10 \times \log_{10}\left(\frac{A}{B}\right)$$

Decibels is usually shortened to dB.

For example, the ratio of the power consumptions of a 100 W and a 60 W light bulb in decibels is

$$\text{ratio of power (dB)} = 10 \times \log_{10}\left(\frac{100}{60}\right) = 2.22 \, \text{dB}$$

So a 100 W light bulb is 2.22 dB more powerful than a 60 W light bulb.

What is the power in an electrical signal? Power is the rate at which energy is changed from one form to another. In the case of light bulbs it is changed from electrical energy to heat and light. The power in an electrical signal is just the same: it is the rate at which the electrical energy in the signal is changed into another form (usually heat). In practice, the absolute power level present in a signal is unimportant so long as it is low enough that electronic components handling the signal do not overheat. Filters and other electronic devices that are used to manipulate signals are described in terms of their effect on signal power, not because overheating is a problem, but because it is a very convenient way of describing their properties. For example, suppose a low-pass filter has a gain of 1 from 0 to 100 Hz, after which it decreases. A 50 Hz signal with a power level of 1 mW applied to the input is passed unchanged by the filter (the gain is 1) so the output is a 50 Hz signal with a power of 1 mW. Now change the frequency to 200 Hz and again apply 1 mW of power to the input. The output of the filter is measured and found to be 0.5 mW (the other 0.5 mW has warmed up the components of the filter by a minuscule amount).

How can the effects of the filter on the 200 Hz signal be described? The usual way is to compare the input and output signals. The gain is the amount by which the filter multiplies the size of the signal going through it. The gain in power is the output power divided by the input power, so for the 200 Hz signal the gain in power is 0.5/1.0 = 0.5. If we multiply the input power (1.0 mW) by the power gain (0.5) we get the output power (0.5 mW). To change the power gain to dB, take the logarithm of the ratio and multiply it by 10:

$$\text{power gain (dB)} = 10 \times \log_{10}\left(\frac{0.5}{1.0}\right) = -3.01 \, \text{dB}$$

Because decibel is always a ratio of powers, we can just say that the gain of the filter at 200 Hz is -3.01 dB. We know that when talking about dB we always mean a ratio of powers.

Calculating the output power of a signal applied to this filter is slightly more difficult if the filter gain is in dB, because the input power cannot just be multiplied by the gain in dB. The gain in dB first has to be converted back to a power

TABLE 3.1 **SOME RATIOS OF POWERS, EXPRESSED AS A RATIO AND IN DECIBELS (dB)**

Power ratio	dB
100	20
2	3.01
1	0
0.5	-3.0
1/10	-10
1/1,000,000	-60

gain, and then the input power is multiplied by this value to obtain the output power. Converting from dB back to a power gain is easy; it is just the reverse of calculating the power gain in dB:

$$\text{power gain} = 10^{(\text{gain in dB}/10)}$$

Working in dB can be a little unusual at first. Table 3.1 shows some examples of power ratios in ordinary units and dB. Note that ratios greater than 1 are positive and those less than 1 are negative in dB.

There is one more very important point with using decibels. Up until now, dB has been defined in terms of ratio of powers. Very often, it is easier to measure voltages than powers yet we would still like to specify ratios in dB because it is so convenient. Decibel always refers to ratio of powers so a method of converting from units of voltage (V) to units of power (W) is needed.

The answer lies with Ohm's law. Ohm's law deals with *resistance*, which is a property of objects that conduct electricity. If a current I is passed through a piece of wire and the voltage V across the ends of the wire is measured, the resistance (R) of the wire is given by Ohm's law:

$$R = \frac{V}{I}$$

The power generated in the wire (in the form of heat) is given by

$$\text{power (W)} = V \times I$$

These two equations can be combined, eliminating the current I, to give

$$\text{power} = \frac{V^2}{R}$$

This equation gives the power generated in a resistor when a voltage is applied across it. For example, it could be used to calculate the resistance of a 60 W light bulb when it is lit up. The power generated in the bulb is 60 W. The voltage across it is the voltage supplied by the power company, which in Canada is 120 V. In the USA it would be 110 V and in most of the rest of the world it is 240 V. In this example 120 V is used. Plugging the numbers into the equation gives

$$60 = \frac{120^2}{R}$$

so R is 240 Ω.

Now imagine that two voltages, V_1 and V_2, are applied to the same resistor R (not simultaneously). The power generated in the resistor when the voltage across it is V_1 is V_1^2/R watts. Likewise, the power generated in the resistor when the voltage across it is V_2 is V_2^2/R watts. The ratio of these powers is

$$\text{ratio of powers} = \frac{V_1^2/R}{V_2^2/R}$$

This is a ratio of powers and so can be converted into decibels:

$$\text{ratio of powers (dB)} = 10\log_{10}\left(\frac{V_1^2/R}{V_2^2/R}\right)$$

and after simplification it becomes

$$\text{ratio of powers (dB)} = 20\log_{10}\left(\frac{V_1}{V_2}\right)$$

Thus if V_1 and V_2 are voltages applied across the same resistor the ratio of their powers in dB is given by

$$\text{ratio of powers (dB)} = 20\log_{10}\left(\frac{V_1}{V_2}\right)$$

The actual value of the resistor does not matter. This is a very useful result. It means that measurements and calculations can be performed on voltages but still use decibels. All that has to be done is to multiply the log of the ratio of voltages by 20, instead of 10.

The key points about decibels are therefore:

1. The decibel is a ratio of powers expressed in logarithmic units.
2. The ratio in dB of two powers, A and B, is given by

$$\text{ratio of powers (dB)} = 10 \times \log_{10}\left(\frac{A}{B}\right)$$

3. The ratio in dB of two voltages V_1 and V_2, applied across the same resistance, is given by

$$\text{ratio of powers (dB)} = 20 \times \log_{10}\left(\frac{V_1}{V_2}\right)$$

In practice, decibels are so useful that the restriction that the resistances be the same is not followed all that rigidly, but that does not matter. Ratios of voltages crop up so frequently in signal processing that the same formula is used to express the ratio in dB of any two voltages, without worrying about whether the resistances are the same in both cases.

Some examples of the use of decibels may be useful:

1. The gain of many types of filters is $1/\sqrt{2}$ at their corner frequency. Thus if a 1 V signal at the corner frequency is applied to the input of the filter, the voltage at the output will be $1/\sqrt{2}$ V or 0.71 V. What is the filter gain in dB at this frequency?

The signals are specified in volts, so the formula for calculating dB directly from voltages is used:

$$\text{gain (dB)} = 20 \times \log_{10}\left(\frac{\text{output voltage}}{\text{input voltage}}\right)$$

$$= 20 \times \log_{10}\left(\frac{1}{\sqrt{2}}\right)$$

$$= -3.01\,\text{dB}$$

The result is negative: if the ratio is less than 1 the result in dB will be negative.

A power gain of 0.5, or -3.01 dB, crops up a lot in filters. The reason is that the corner frequency of many types of filters is defined as the point at which the filter output power is half the input power. That is why, in Figure 3.10, the corner frequency is marked as the point at which the gain has fallen to $1/\sqrt{2}$ because this is when the output power is half the input power. Not all filters follow this convention but many do.

Since $-3.01\,\text{dB}$ is so close to $-3.00\,\text{dB}$, the latter is usually used, especially when we are being descriptive rather than making precise calculations. So the corner frequency of filters is often denoted by $f_{-3\,\text{dB}}$ rather than f_c.

2. The gain of the low-pass filter in Figure 3.10 is 1 at low frequencies. What is this gain in dB?

Once again, the formula for calculating dB directly from voltages is used:

$$\text{gain (dB)} = 20 \times \log_{10}\left(\frac{\text{output voltage}}{\text{input voltage}}\right)$$

$$= 20 \times \log_{10}(1)$$

$$= 0\,\text{dB}$$

So a gain of 0 dB does not mean that nothing gets through the filter: it means that the signal passes through without being altered at all.

3. Consider filter (b) in Figure 3.11. The manufacturer specifies that the gain in the stopband (where ideally no signal should get through) is $-80\,\text{dB}$. If a 100 mV signal at a frequency that is in the stopband is applied to the input of the filter, what is the output voltage?

The equation for calculating gain from voltage ratios needs to be reversed:

$$\text{gain (dB)} = 20 \times \log_{10}\left(\frac{\text{output voltage}}{\text{input voltage}}\right)$$

can be re-arranged to give

$$\frac{\text{output voltage}}{\text{input voltage}} = 10^{(\text{gain (dB)}/20)}$$

Now plug in the numbers:

$$\frac{\text{output voltage}}{\text{input voltage}} = 10^{(-80/20)}$$

$$= \frac{1}{10{,}000}$$

So if the input voltage is 100 mV the output voltage is 10 μV.

4. What is a gain of 0 in dB?

Gain is defined in terms of voltage, so the formula for calculating dB directly from voltages is used:

$$\text{gain (dB)} = 20 \times \log_{10}\left(\frac{\text{output voltage}}{\text{input voltage}}\right)$$

$$= 20 \times \log_{10}(0)$$

This presents a problem, because the logarithm of 0 is not defined. But in real life this does not occur because it is not possible to build a filter with a gain of 0, that is one that totally blocks a signal. All that has to be done is to reduce the unwanted parts of a signal to a low enough level that they become insignificant. We can calculate just how low that level needs to be and then specify the filter gain that we need to achieve it. In practice, filter gains usually do not go much below $-80\,\text{dB}$ to $-100\,\text{dB}$ but for many purposes this may be overkill and a filter with a stopband gain of $-50\,\text{dB}$ may be adequate, and cheaper than one with $-100\,\text{dB}$ of stopband gain.

5. If the gain of a filter in its stopband is $-80\,\text{dB}$, what is the corresponding attenuation?

The gain in dB is calculated from

$$\text{gain (dB)} = 20 \times \log_{10}(\text{gain})$$

Earlier, attenuation was defined as (1/gain) so substitution yields

$$\text{attenuation (dB)} = 20 \times \log_{10}\left(\frac{1}{\text{gain}}\right)$$

But $\log(1/x)$ is equal to $-\log(x)$, so

$$\text{attenuation (dB)} = -20 \times \log_{10}(\text{gain})$$

$$= -(-80\,\text{dB})$$

$$= 80\,\text{dB}$$

In general, if the gain is $x\,\text{dB}$, the attenuation is $-x\,\text{dB}$.

With the basics of working in decibels covered, we return to filters. Look at Figure 3.11. The steepness of the gain–frequency graph in the middle of the transition band is called the roll-off rate. This is an important property of a low-pass filter because it determines how well a filter will separate two signals of different frequencies. In general, the roll-off rate is not a linear function of frequency so it cannot be specified in terms such as the gain falls by 0.1 every 10 Hz. It turns out, however, that the roll-off rate for many types of filters is linear if the gain–frequency graph is plotted as a log–log plot. This is what has been done in Figure 3.11: log(gain) has been plotted against log(frequency) The roll-off rate is constant in the middle of the transition band so the graph is a straight line in this region. A suitable way of quantifying the roll-off rate, that is, the slope of the graph, is needed bearing in mind that the graph is in log–log units. A common way of doing this is to express the roll-off rate in *decibels per octave*. Specifying gain in dB takes care of the gain axis, because dB is a logarithmic scale. Suitable units for log(frequency) are now needed, which is where octaves come in.

If a frequency is increased by one octave, the frequency is doubled. This means that if changes in frequency are expressed in octaves, rather than Hz, they are converted to a logarithmic scale. Writing this a little more mathematically may make things clearer. Take two frequencies, f_1 and f_2. Their ratio is f_1/f_2. This is converted to octaves using the relationship

$$\text{frequency ratio (octaves)} = \log_2\left(\frac{f_1}{f_2}\right)$$

Using logs to the base 10 rather than base 2 gives the relationship

$$\text{frequency ratio (octaves)} = \frac{\log_{10}(f_1/f_2)}{\log_{10}(2)}$$

$$= 3.3219 \times \log_{10}\left(\frac{f_1}{f_2}\right)$$

Like dB, octaves are a scale of frequency ratios rather than absolute values. Octaves are used frequently in music. A good singer might have a range of 3.5 octaves, meaning that the highest note that she/he can sing has a frequency 3.5 octaves higher than the lowest note that she/he can sing. Describing the singer's range in terms of octaves does not give any information about the values of the lowest and highest frequencies in Hz. The singer can be a soprano or a tenor and still have a range of 3.5 octaves.

By expressing both the gain and frequency in logarithmic units, the roll-off rate, which is often constant in the middle of the transition band, is easy to specify. A typical value might be 48 dB/octave. This means that the gain of the filter decreases by 48 dB each time the frequency increases by one octave (i.e. each time the frequency doubles).

At the start and end of the transition band the gain–frequency graph (expressed in log–log units) is often not linear. There is no easy way of specifying the gain of the filter at these frequencies. If accurate calculations in this part of the filter's frequency response are needed, there are a couple of options. If the filter is of a standard design (see the next section) the gain of the filter may be expressible in analytical form as an equation. Alternatively, the gain can be read from a graph provided by the manufacturers. Usually the performance of the filter near the corner frequency is not that important and in this case a simplified version of the gain–frequency graph can be used on low-pass filters of type (a) in Figure 3.11, that is, filters whose gain just keeps on falling after the corner frequency. It is assumed that the filter gain is 1 (0 dB) up to the corner frequency and then decreases straight away at the roll-off rate. Figure 3.12 shows the real and simplified response of this type of low-pass filter. As long as accurate calculations at the corner frequency are not required, this simplification is often good enough to work with.

As an example, imagine that a signal from a pneumotachograph, representing respiratory flow, is being recorded. The flow signals are being recorded from subjects at rest and it is likely that there is not much useful information in the signal above 5 Hz. Mixed in with the respiratory signal is some interference at 60 Hz, which will be removed by passing the flow signal through a low-pass filter with a corner frequency of 10 Hz and a roll-off rate of 24 dB/octave. How much will the 60 Hz interference be attenuated by the filter?

To tackle this problem the simplified model of the low-pass filter's amplitude response in Figure 3.12 is used. This model has a constant gain of 1 (0 dB) up to 10 Hz, at which point it decreases at 24 dB/octave. The position of the 60 Hz signal in the transition band is calculated next. The transition band starts at 10 Hz and we need to know how many octaves higher 60 Hz is compared to 10 Hz. The ratio of the frequencies is 60/10, so using the formula given earlier:

$$\text{no. octaves} = 3.3219 \times \log_{10}\left(\frac{f_1}{f_2}\right)$$

$$= 3.3219 \times \log_{10}\left(\frac{60}{10}\right)$$

$$= 2.58 \text{ octaves}$$

The roll-off rate is 24 dB/octave, so in 2.58 octaves the gain decreases by $2.58 \times 48 = 61.9$ dB. The gain at the start of the transition band was 0 dB so at 60 Hz it is $0 - 61.9$ dB $= -61.9$ dB.

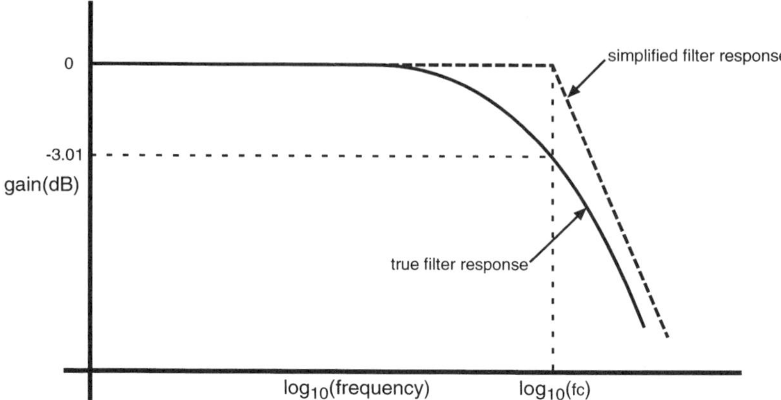

Figure 3.12 The true and simplified frequency response of a low-pass filter. In the simplified response the gain is assumed to be 1 up to the corner frequency, at which point it decreases at the roll-off rate.

This gain in dB is converted back to a ratio of voltages using the formula

$$\text{voltage ratio} = 10^{(\text{gain(dB)}/20)}$$

$$= 10^{(-61.9/20)}$$

$$= \frac{1}{1,245}$$

If the original 60 Hz signal has an amplitude of 200 mV then the filter will reduce this to $200/1,245 = 0.16$ mV.

TYPES OF LOW-PASS FILTERS

There are several different ways of designing and making filters. There are three 'classic' filter types: Butterworth, Chebyshev (the spelling of the latter varies a little) and Bessel. Their performance (amplitude response, etc.) is based on theory and is specified mathematically. Over the years, a number of different ways of making real filters that perform in the same way as predicted by theory have been developed. It is not necessary to know the exact circuit configuration used to make a filter to know how it will perform: all one needs to know is the filter type (Butterworth, etc.).

There is no ideal filter. Each type of filter is good in some areas but poor in others, and filter selection is a matter of choosing the best one for a particular

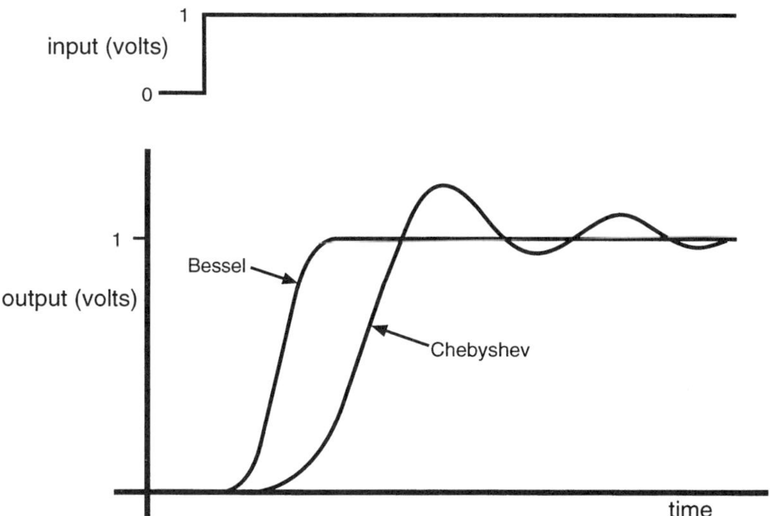

Figure 3.13 This graph shows the output voltage against time when the input voltage to two low-pass filters is suddenly changed from 0 to 1 V. The output of the Bessel filter rises smoothly to 1 V, whereas the output of the Chebyshev filter oscillates about the 1 V level before settling down.

purpose. Of the classic filter types, Butterworth filters have the flattest passband but a poor roll-off rate. Chebyshev filters have some passband ripple but a better (steeper) roll-off rate. Bessel filters have the worst roll-off rate of the three. They are used in some applications because they have the best performance in another area: the *phase* response. All filters delay the signals that pass through them to some extent: somewhat like recording the signal on a tape recorder and then playing it back a short time later. This delay is not normally a problem so long as it affects the whole signal evenly. The phase response of a filter describes how the delay in the filter changes with the frequency of the signal passing through it. If the filter delays some frequencies more than others it will distort the signal as it passes through the filter. A good phase response also means that the filter will respond in a predictable way to sudden changes in the signal level, such as spikes and steps. The Chebyshev filter has a poor phase response and thus does odd things when given a sudden change in signal level. Figure 3.13 shows what happens when a step change in signal is applied to a Chebyshev filter – the output 'wobbles' or 'rings' slightly before settling down to the new value. In contrast, the Bessel filter (which has a better phase response) causes minimal distortion of the step.

The terms *order* and *pole* are also used to describe filters. Going back to classic filter design again, these filters are built in stages and the more complex filters are made by stacking several stages together. A tremendous amount of research has gone into designing filter circuits. The most common type of circuit used (at the frequencies we are interested in) is the *active filter*, which is made up from resistors, capacitors and an amplifier (the 'active' part). Each stage of an active filter contains one amplifier and a collection of resistors and capacitors. The resistor and capacitor values used set the type of filter and its corner frequency. Horowitz and Hill (pp. 266–84) have an excellent section on active-filter design. In practically all active-filter designs each stage has two poles, so a filter with two stages is called a 'four-pole' or 'fourth-order' filter.[2] One of the useful things about this classification system is that the maximum roll-off rate is 6 dB per octave per pole, so that a fourth-order filter will have a maximum roll-off rate of 24 dB/octave. This applies to all three classic filter types. As the number of poles goes up so does the price, and eighth-order filters are about the most complex ones available 'off the shelf' at an affordable price.[3] The '6 dB per octave per pole' rule does vary slightly, especially near the corner frequency, so check the manufacturer's data if the filter performance is critical.

Low-pass filters are not the whole answer to preventing aliasing. There are limits to their performance, they may alter the signal undesirably and they can be quite costly. Also, one filter is needed for each channel so for a typical data acquisition system sixteen filters are required. If signals recorded on different channels are going to be compared to one another then *matched* filters may be necessary. Matched filters have a performance (amplitude response, etc.) that is guaranteed by the manufacturers to be the same to some tolerance such as ±1%. If matched filters are not used, what appears to be a difference between two signals may only be a difference in filter properties.

OVERSAMPLING

The other way to prevent aliasing is simply to increase the sampling rate so that it is well above twice the frequency of any signal components that may reasonably be expected to be present. Raising the sampling rate above the minimum required is called *oversampling*.

The penalty for oversampling is increased storage space requirements for all the numbers and increased processing time because every calculation (e.g. finding the mean) involves more numbers. The tide is swinging away from the use of steep (high roll-off rate) anti-aliasing filters towards higher sampling frequencies because the prices of computers, fast ADCs and data storage are all falling rapidly. However, it is still easy to collect so much data that computers and software choke on them. This is particularly important if real-time data analysis is used. Sixteen channels of data collected on a 12-bit ADC running at 1 kHz is 16,000 samples and as each 12-bit sample takes up two bytes of storage space the data occupy 31 kbyte of memory per second. In 33 s there will be 1 Mbyte of data, and 110 Mbyte of data after an hour. Storing this amount of data is not too difficult because writeable CDs store 650 Mbyte each which is 5.9 hours of data. On the other hand, there is only 63 µs to process each sample before the next one arrives. Even with fast computers this is not much time to perform calculations if real-time analysis is desired. Very commonly, the two techniques for preventing aliasing discussed above (low-pass filters and oversampling) are combined. This relaxes the specification of the filter, making it cheaper, whilst keeping memory and processing requirements under control.

PREVENTING ALIASING

Here are some ideas on simple methods for choosing a suitable combination of anti-aliasing filter and sampling rate:

1. Estimate the frequency spectrum of the signal to be recorded. Some suggestions on how to do this are given later in this chapter and in Chapter 5.
2. Choose a frequency below which most (e.g. 95–99%) of the signal is present. As an example, suppose that there is very little useful signal beyond 30 Hz. Call this frequency f_{max}, so in this example $f_{max} = 30$ Hz.
3. Select a low-pass filter with a corner frequency 1.5–2 times higher than f_{max}. In this example the corner frequency should therefore be 45–60 Hz. If the corner frequency is set to be the same as f_{max}, some of the high-frequency components of the signal may be lost, because at the corner frequency the gain of the filter is less than 1.
4. Decide on the Nyquist frequency. A typical value is 2–3 times the filter corner frequency using a filter with a good roll-off rate (greater than or equal to 24 dB/octave or 4 poles). For this example the Nyquist frequency is 90–120 Hz.
5. Sample at twice the Nyquist frequency, that is, 180–240 Hz in this example.

This approach is quite conservative, because if f_{max} is 30 Hz then the lowest theoretical sampling rate is 60 Hz. On the other hand, it gives a system that works with minimal mathematical effort.

FINDING THE MAXIMUM FREQUENCY PRESENT IN A SIGNAL

The most important question when sampling a real-world signal is 'What is the maximum frequency present in the signal to be recorded?' There are several ways of finding this out accurately. The best way is to measure it with a *spectrum analyser*, a piece of electronic equipment that plots the amplitude of all the frequencies present, but spectrum analysers are expensive and not always available. A data acquisition/analysis system can be used to do the same thing and Chapter 5 goes through some of the techniques that are involved.

It is useful, however, to have some rough-and-ready ways of estimating the power spectra of signals, and in particular the frequency f_{max} above which there is very little signal power, so that a reasonable initial combination of sampling rate and filter can be selected. The first thing to do is to get some idea of what the signal looks like, that is, obtain a plot of the signal amplitude against time. An oscilloscope is ideal for this. An alternative is to record the signal on a data acquisition/ analysis system with the sampling rate set to a high value such as 10 kHz. This high rate is used so the signal is heavily oversampled to avoid aliasing. Look at the signal to see what features it has:

1. Does the signal have some regular repeating pattern to it (i.e. is it *periodic*)?
2. Is the signal smooth and roughly similar to a sine wave (see Figure 3.16), or rough and irregular?
3. Are there any narrow spikes or sharp edges?

Examples of periodic signals are arterial blood pressure, respiratory flow and ECG recordings (Figure 3.14).[4] In contrast, signals such as EMGs and evoked potentials (Figure 3.15) are nonperiodic.[5] With periodic signals it is relatively easy to estimate f_{max}. At the start of this chapter it was mentioned that any signal can be made up by adding together a number of sine waves. A sine wave is a 'pure' signal, that is, it only has a single frequency component. Figure 3.16 shows a recording of a sine wave. Its frequency can be specified exactly and measured in a variety of ways. The simplest is to measure the time between two points on adjacent waves. This time is the *period*, and is related to the frequency:

$$\text{frequency (Hz)} = \frac{1}{\text{period (s)}}$$

The sine wave in Figure 3.16 has a period of 0.2 s so the frequency is 5 Hz. Any two corresponding points can be used to find the period but the easiest ones to measure from are the peaks and the points where the signal crosses zero.

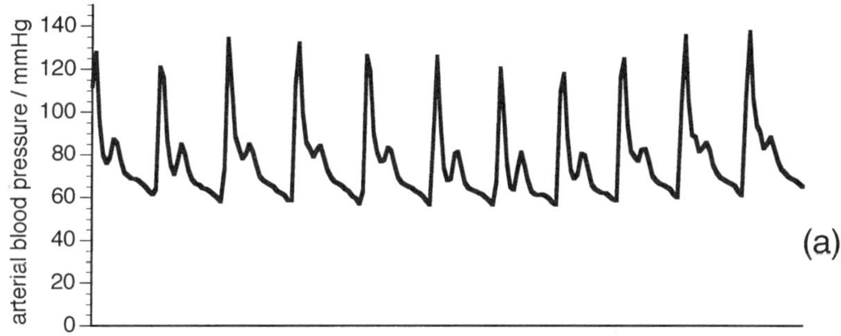

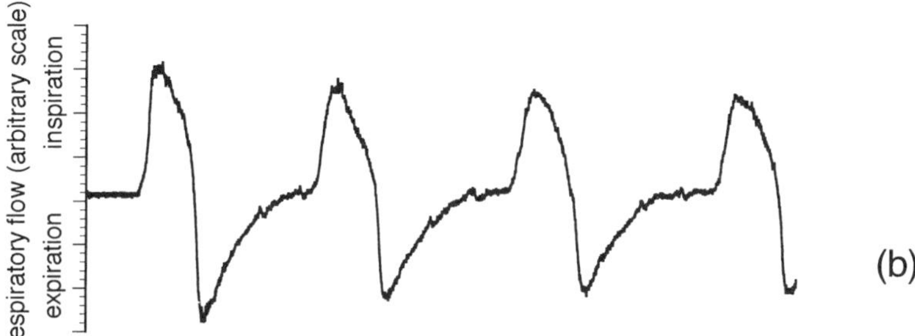

Figure 3.14 Examples of periodic signals: (a) is a recording of arterial blood pressure and (b) is a recording of respiratory airflow.

Figure 3.15 An example of a non-periodic signal. The graph shows four superimposed magnetic motor evoked potentials (MMEPs). The MMEP is a recording of the electrical activity of a muscle when the motor cortex is stimulated noninvasively by a rapidly changing magnetic field. These recordings were made from the anterior tibial muscle of a dog. The spike at time zero is an artefact produced by the magnetic stimulator.

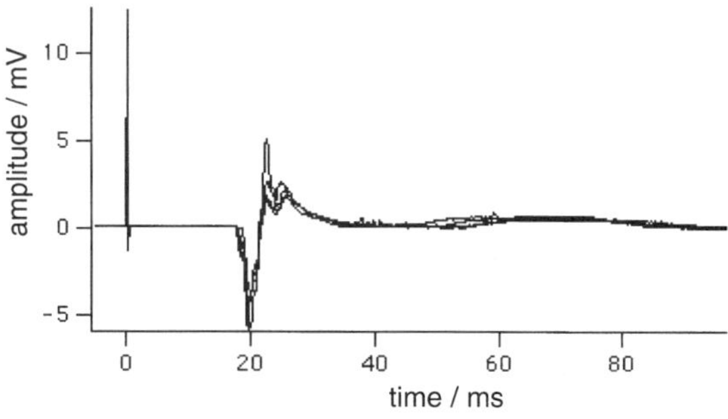

Figure 3.17 is a relatively smooth, periodic signal that is not very different from an arterial blood pressure recording. It consists of just three sine waves added together, with frequencies of 1 Hz, 2 Hz, and 3 Hz and different amplitudes and phases. The fact that the frequencies are integer multiples of each other is not just convenience but is the application of an important theorem. Fourier's theorem

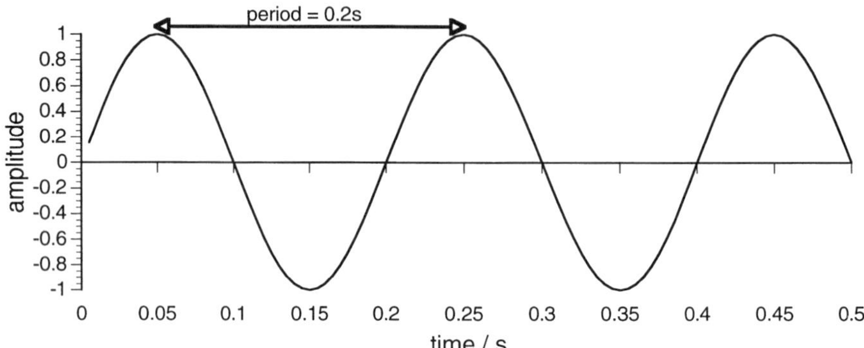

Figure 3.16 A sine wave. The period is the time between two identical points on adjacent cycles.

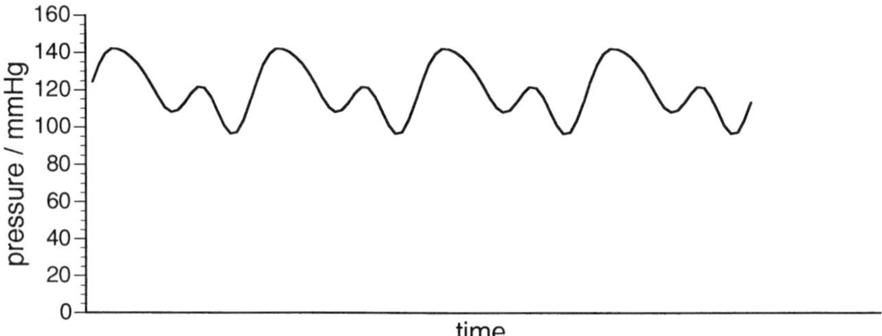

Figure 3.17 An artificial 'blood pressure' recording made up from the sum of three sine waves. The three sine waves have the following properties: (1) frequency 1 Hz, amplitude 15.0, phase angle 0°; (2) frequency 2 Hz, amplitude 12.0, phase angle $-47°$; (3) frequency 3 Hz, amplitude 4.7, phase angle $+30°$. The final trace is made up of the sum of these waves plus the mean value of 120.

states that any *periodic* signal can be made up from a number of sine waves of different amplitudes but whose frequencies are integer multiples of each other. The lowest frequency is called the *fundamental frequency* or just the *fundamental*, and the higher frequencies are called *harmonics*. The sine wave whose frequency is three times higher than the fundamental, for example, is called the third harmonic.

The square wave in Figure 3.18a is also a periodic signal, so it can be constructed (*synthesized*) by adding together sine waves whose frequencies are integer multiples of the fundamental frequency (f_0). It can be shown theoretically that a square wave is made up of the fundamental frequency plus just the odd harmonics in decreasing amplitude:

$$\text{square wave} = f_0 + \frac{f_3}{3} + \frac{f_5}{5} + \frac{f_7}{7} + \cdots$$

where f_0 represents the fundamental sine wave, f_3 is the third harmonic ($3 \times f_0$), and so forth.

In Figures 3.18b–e the square wave shown in (a) is gradually synthesized by adding in harmonics of higher and higher frequency. Figure 3.18b is the fundamental frequency f_0 plus the third harmonic f_3. Figure 3.18c has the fundamental, third and fifth harmonic. The last waveform in Figure 3.18e goes up to the fifteenth harmonic. As more harmonics are added the signal becomes more like a square wave.

There are two important concepts that are illustrated in Figure 3.18. Firstly, the more harmonics that are added the less the signal looks like a sine wave. Secondly, the higher harmonics are responsible for defining the parts of the waveform that change quickly. The first seven or so harmonics define the rough shape of the square wave and the higher ones sharpen up the rising and falling edges and smooth out the ripples on the plateaus.

The reverse is also true and forms two useful rules of thumb:

1. The more a periodic signal looks like a sine wave, the less harmonics it will have.
2. A periodic signal that changes level quickly, that is, one that has sudden steps, steep sections or spikes, will have a large number of harmonics.

A perfect square wave (one that takes no time to change from a low to a high voltage) contains an infinite number of harmonics. Real-world square waves, however, take a small amount of time to change from a low to a positive high and so the number of harmonics that they contain is large but finite.

The fundamental frequency of any periodic signal is found with the same technique that was used for a sine wave. The time between corresponding points on adjacent waves is the period, and the reciprocal is the fundamental frequency. For example, in the recording of respiratory airflow in Figure 3.19 the time between successive breaths is 4.1 s so the fundamental frequency is 0.244 Hz. This is, of course, the same as the respiratory frequency. In a similar fashion the fundamental frequency of an ECG or blood pressure trace is the same as the heart rate.

Having worked out the fundamental frequency of our periodic signal, the next step is to decide how many harmonics are contributing significantly to the signal. Use one of the rules of thumb: the more the signal looks like a sine wave and the smoother it is the less harmonics it will contain. Figure 3.20a is a recording of

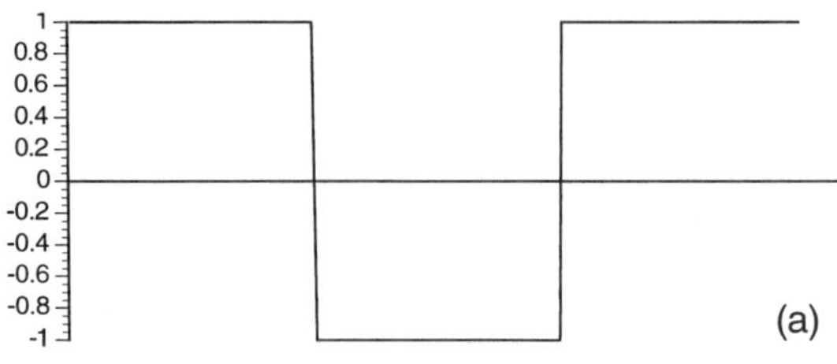

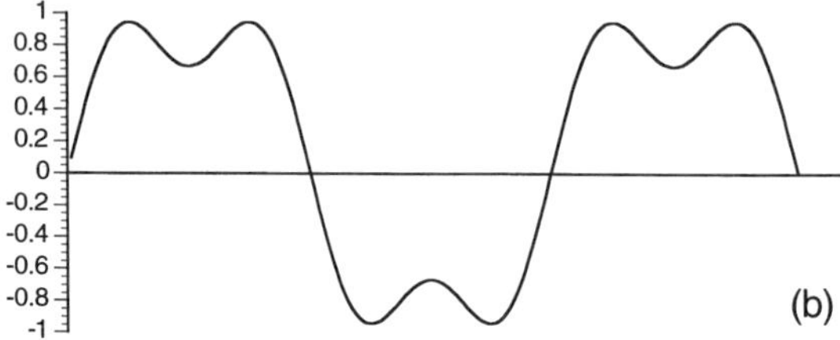

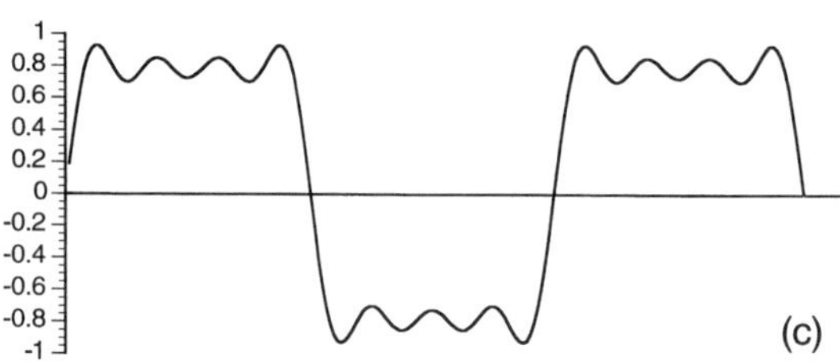

Figure 3.18 Synthesis of a square wave from sine waves: (a) a square wave with fundamental frequency f_0; (b) fundamental sine wave plus the third harmonic $(f_0 + f_3/3)$; (c) fundamental, third and fifth harmonics $(f_0 + f_3/3 + f_5/5)$; (d) fundamental and third to seventh harmonics $(f_0 + f_3/3 + f_5/5 + f_7/7)$; (e) fundamental and third to fifteenth harmonics $(f_0 + f_3/3 + f_5/5 + f_7/7 + f_9/9 + f_{11}/11 + f_{13}/13 + f_{15}/15)$.

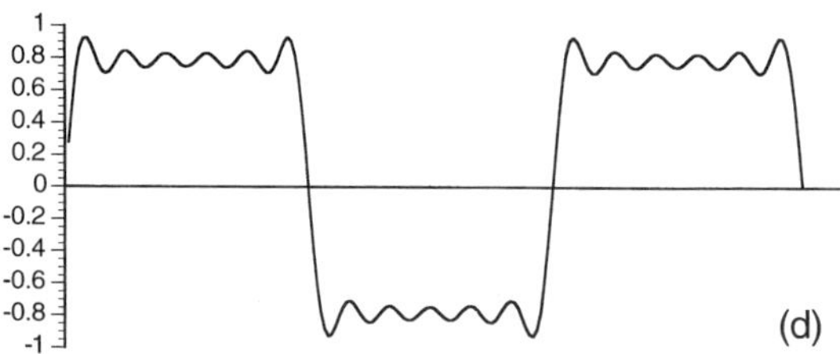

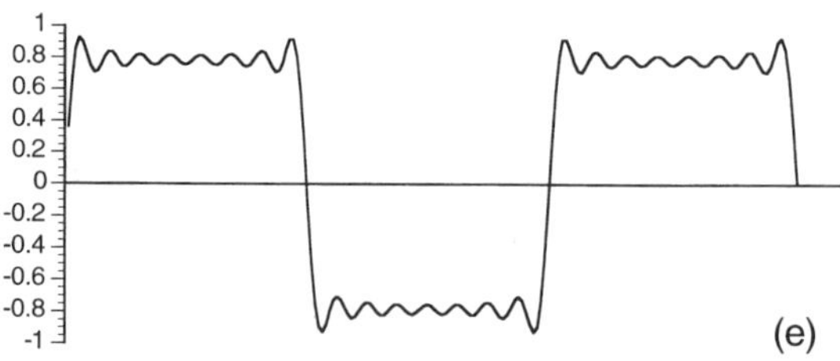

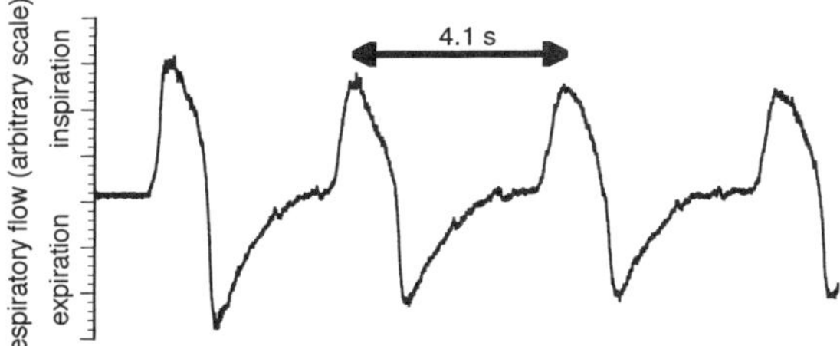

Figure 3.19 A recording of respiratory airflow from a resting human. The period is 4.1 s.

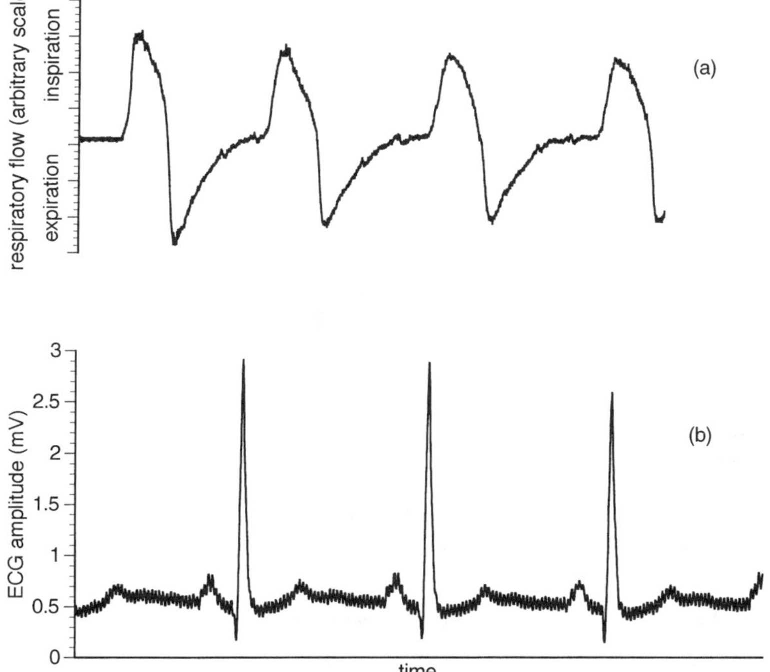

Figure 3.20 (a) A recording of respiratory airflow from a resting human. (b) The ECG from a human subject. The period is the time between the large R waves and is 0.833 s. The small oscillations are 60 Hz interference.

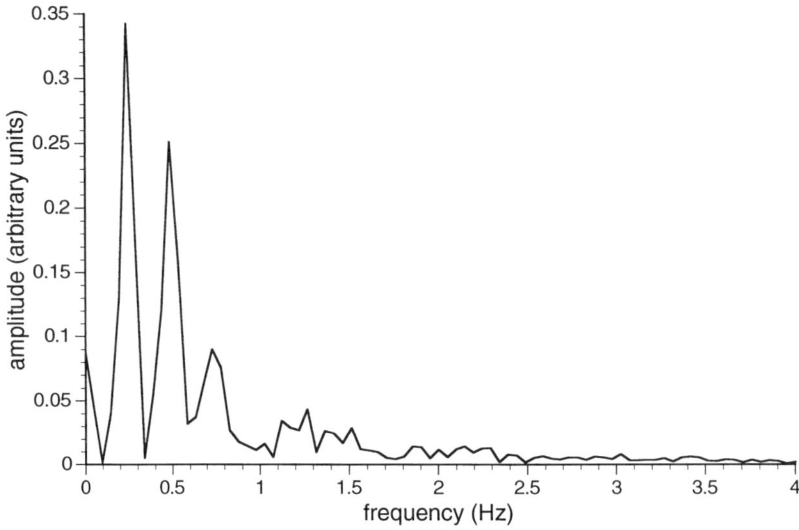

Figure 3.21 The amplitude spectrum of the respiratory flow signal shown in Figure 3.20a. The three large spikes at 0.244 Hz, 0.488 Hz, and 0.732 Hz are the fundamental frequency and the second and third harmonics.

respiratory airflow from a resting human subject. It is relatively smooth with no sharp edges or spikes (there is a small amount of noise present as well which makes the tracing slightly fuzzy) so one would expect the signal to have few harmonics, perhaps four or five. The time between successive breaths is 4.1 s and thus the fundamental frequency is 0.244 Hz. As a rough estimate, therefore, most of the flow signal will be made up of a fundamental sine wave at 0.244 Hz plus four or five harmonics. The second harmonic will be at 0.488 Hz, the third at 0.732 Hz, the fourth at 0.976 Hz, and the fifth at 1.22 Hz (the first harmonic is the same as the fundamental frequency). The spectrum of the flow signal (a plot of the amplitudes of the sine waves present in the signal against their frequency) is shown in Figure 3.21. The fundamental frequency and the second and third harmonics are clear. There does not appear to be much of a fourth harmonic, the fifth and sixth can be made out and after that there is little signal present. One can state with confidence that there is virtually no signal present above 2.44 Hz (the tenth harmonic) so f_{max} for this respiratory flow signal is about 2.5 Hz.

In contrast the ECG signal in Figure 3.20b is not at all like a sine wave: there are flat portions and narrow spikes so it will contain many harmonics, about 50. The fundamental frequency is 1.2 Hz (72 beats/min) so the fiftieth harmonic is 60 Hz. Figure 3.22 shows the spectrum of this signal. The spike at 0 Hz is unimportant: it is caused by the fact that the average value of the signal is not 0. There is a large

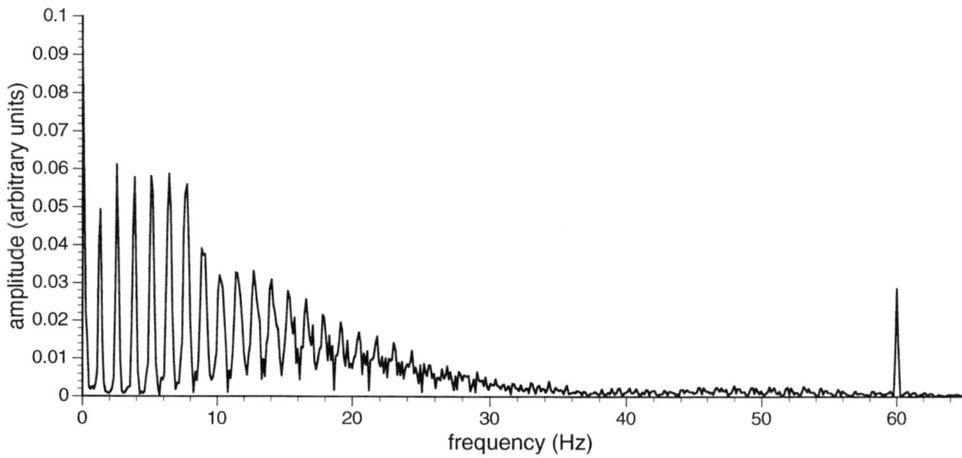

Figure 3.22 The amplitude spectrum of the ECG signal shown in Figure 3.20b. Harmonics of significant amplitude are present to about 55 Hz, and the 60 Hz interference is clearly visible.

component at the fundamental frequency of 1.2 Hz. The next five harmonics (2.4 Hz to 7.2 Hz) are all large and of about the same amplitude. After that the harmonics decrease in amplitude and fade away at around 40 Hz, but reappear at a low level between 45 Hz and 55 Hz. Above 55 Hz there is very little signal except for a spike at 60 Hz. The spike at 60 Hz is interference from the electrical supply to the building which can be seen on the original signal (Figure 3.20b) as a small sawtooth-like pattern superimposed on the ECG waveform. Above 60 Hz there is little signal, so f_{max} is around 65 Hz.

Judging the 'spikyness' of a signal in this manner is not intended to take the place of more precise measurements but it is still very useful. One thing to remember is that sampling at a higher rate than is strictly necessary is always safe – the only penalty is more numbers to deal with.

Nonperiodic signals are still made up of sine waves, but the frequencies of the sine waves are no longer integer multiples of each other and can be of any frequency. There is an upper limit (f_{max}) to the frequencies of the sine waves that make them up, but it is not easy to estimate f_{max} by looking at the signal. The 'spiky' rule of thumb still applies: nonperiodic signals that change level suddenly have a high f_{max}.

In general, nonperiodic signals require higher sampling rates than periodic signals. When dealing with nonperiodic signals it is best to set the sampling rate initially so that there are enough samples to follow the signal closely. This approach leads to a sampling rate that is higher than necessary but at least the signal is not aliased. The spectrum can then be calculated and an estimate of f_{max}

obtained. The sampling rate can then be reduced if necessary. The sampling theorem applies to all signals, whether periodic or not: to avoid aliasing the signal must be sampled at a rate at least twice as high as f_{max}.

Another approach to estimating f_{max} can be used in some circumstances. In many cases the data acquisition system will be recording the output of an amplifier or some type of analyser. There will be an upper limit to the frequency response of the amplifier or analyser that we can call f_{hi}. If we know f_{hi} we can estimate f_{max} since, no matter what the input to the amplifier or analyser does, the output will not contain frequencies much above f_{hi}. If you are lucky the frequency response of the equipment will be given in the manufacturer's specifications. For example, the ECG amplifier in our Datex Cardiocap II anaesthesia monitor has a 3 dB bandwidth of 0.5–30 Hz (i.e. $f_{hi} = 30$ Hz), and the invasive blood pressure channel has a 3 dB bandwidth of 0–20 Hz. What the manufacturers almost certainly will not give is the roll-off rate, so that even though you know f_{hi} you do not know how rapidly the output falls to negligible levels above this. There are two things that you can do. Firstly, since you know that the output of the amplifier is unreliable above f_{hi}, you can follow the amplifier with a low pass filter that has a flat passband to f_{hi} and a steep roll-off above this. Since the gain of the filter at its corner frequency is -3 dB we do not want to set the corner frequency to be the same as f_{hi} because this will not give us a flat passband up to f_{hi}. However, a corner frequency $(1.5–2) \times f_{hi}$ (depending upon the type of filter) will be fine. For our Cardiocap we can make the corner frequency quite close to f_{hi} because we know that the response of the amplifier is already falling off at f_{hi}, and so it is not important that the filter response is flat all the way out to f_{hi}. The other approach is to assume that the amplifier or analyser has a modest roll-off rate of 12 dB/ octave, equivalent to a second-order filter. At $4 \times f_{hi}$ the output will be 48 dB lower than in the passband, which makes it reasonable to call this frequency f_{max}.

In many cases the frequency response of the equipment is not specified by the manufacturer. Instead, the *rise time* may be quoted. This is typical for equipment such as gas analysers. The rise time is how long it takes the output of the analyser to respond to a sudden change in gas concentration. The rise time is usually quoted as the time it takes the output to change from 10% to 90% of its final value. For example, suppose the equipment is a CO_2 analyser. If the concentration of CO_2 at the analyser input is suddenly changed from 0% to 5% (a 'step' change) the analyser output will also change from reading 0% to 5% but it will take a finite time to do this. The way in which the output of the analyser changes in response to a step change at its input is called its *step response*. The rise time in this example is how long it takes the analyser output to change from reading 0.5% to 4.5% CO_2. The rise time is quoted over the range 10%–90% and not 0%–100%

because the output tends to approach its final value exponentially and thus it is difficult to define when the output has settled. On the other hand, the 10% and 90% points are easy to define and measure.

We know that the high-frequency components of a signal are responsible for the rapidly changing sections. Since the rise time represents the maximum rate at which the analyser output can change, it is dominated by the high-frequency response of the analyser. It would be very nice if we could calculate the analyser's high-frequency response in some way from the rise time because this would enable us to estimate $f_{\max}$. This can be done if we know the step response of the analyser. The entire frequency response of a piece of equipment can be calculated from its step response, although the process requires some advanced signal processing. The problem is that we have to know what the whole step response is, that is, we have to sample the step response at a high rate so that we know its shape accurately. This approach will not work if all we know is the rise time because we know nothing about the shape of the step reponse. All we know is how long it takes the output to change from 10% to 90% of its final value.

We can find a solution to the problem if we are prepared to make some assumptions. In general, the frequency response cannot be calculated from the rise time because we do not know the exact shape of the step response. But if we assume that the analyser behaves like a low-pass filter, we know what the step response is. Furthermore, being a filter, we know what the maximum frequency is that the filter will pass because we know its corner frequency f_c and thus we can calculate the relationship between rise time and f_c. If you do the calculations you end up with the following result. The rise time (time taken for the output to change from 10% to 90% of the final value) and corner frequency f_c of a low-pass filter are related by the equation:

$$f_c \cong \frac{0.4}{\text{rise time}}$$

This equation is valid so long as the filter does not have much overshoot in its step response, so it is valid for Butterworth and Bessel but not Chebyshev filters. It is also reasonably valid for analysers because overshoot is an undesirable feature in an analyser's step response and manufacturers avoid it. It also does not matter what the order of the filter is, the value of 0.4 does not change much if the filter order is changed from 2 to 6.

Now that we know the corner frequency f_c of the filter that is equivalent to the analyser we proceed as we did before. We do not know what the roll-off rate of the filter is but it will probably be satisfactory if we assume it is 12 dB/octave. At four times the corner frequency the attenuation is 48 dB, which is not bad and we can make it our Nyquist frequency. Thus the sampling rate that we need to use is $8 \times f_c$. This approach is fairly conservative since we may well not be using the full bandwidth of the analyser.

As an example, a quick look at the specification sheets for some of the equipment in the laboratory here yielded the following:

1. A Datex Cardiocap II CO_2 analyser with a rise time of 270 ms.
2. A Marquette MG-100 mass spectrometer with a rise time of 100 ms.
3. An Axon Instruments patch clamp headstage amplifier with a rise time of 20 μs.

Using the equation above, the CO_2 analyser has an equivalent f_c of $0.4/0.27 = 1.5$ Hz. Sampling at 12 Hz or above will avoid aliasing no matter what gas waveform the analyser measures. Similarly, the mass spectrometer has an equivalent f_c of $0.4/0.1 = 4$ Hz and the patch clamp amplifier has an equivalent f_c of $0.4/(20 \times 10^{-6}) = 20$ kHz. It may seem excessive to require a sampling rate of 8×20 kHz $= 160$ kHz for the patch clamp amplifier, but consider the following. The sampling rate $f_s = 8 \times f_c$. But $f_c \cong 0.4/$rise time so $f_s \cong 3.2/$rise time. The time δt between each sample is $1/f_s$ so $\delta t =$ rise time$/3.2$ and therefore rise time $= 3.2 \times \delta t$. Thus there are only about three samples during the rise time, and those three samples have to fully characterize the shape of the step response in this region.

It was mentioned earlier that sampled signals are not the same as the original continuous signal. This difference becomes greater as the sampling rate approaches the Nyquist limit. Figure 3.23 shows one cycle of a sine wave that has been sampled at 10 (Figure 3.23a) and 2.2 (Figure 3.23b) times its frequency. A sine wave is a pure tone and both sampling rates are greater than twice the frequency of the sine wave so there is no aliasing, yet Figure 3.23b looks nothing like a sine wave. The 'spiky' rule suggests that the signal in Figure 3.23b has many higher-frequency components, which is indeed the case. Figure 3.7 shows that a sampled signal contains the original signal spectrum plus multiple copies of it shifted higher in frequency. As the sampling rate is reduced towards the Nyquist frequency, the extra copies of the spectrum move down in frequency and are packed more closely together, so the signal looks less and less like the original. The plot of the sine wave sampled at ten times its frequency (Figure 3.23a) looks a lot more like a sine wave. This leads to another useful rule of thumb: if the sampled signal looks like the original it is probably being sampled fast enough to prevent aliasing. There is a caveat to this rule, however. Figure 3.24a is a sine wave sampled at twenty times its frequency. Figure 3.24b is the same sine wave

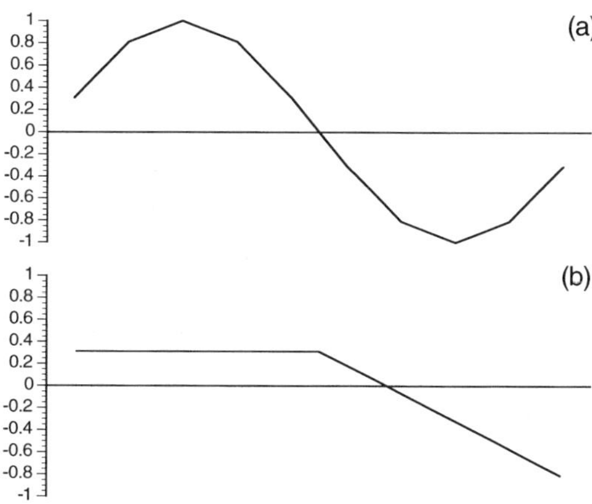

Figure 3.23 (a) A sine wave sampled at ten times its frequency. The shape of the original signal is clear. (b) The same sine wave sampled at 2.2 times its frequency. The sampled signal is quite different from the original, although there is no aliasing and the original sine wave can be recovered intact.

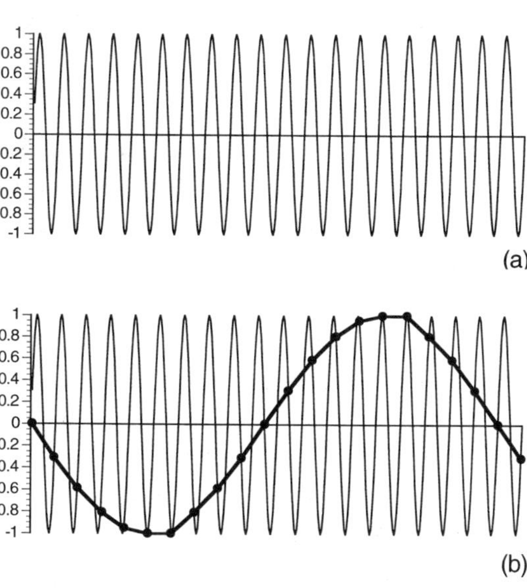

Figure 3.24 (a) A sine wave sampled at twenty times its frequency. The sampled signal closely resembles the original sine wave. (b) The same sine wave sampled at 1.9 times its frequency (heavy line) with the original sine wave superimposed (fine line). Each circle marks the point at which the original signal was sampled. The sampled signal is aliased and resembles a sine wave of much lower frequency.

sampled at 1.9 times its frequency so it has been aliased, but it still looks like a sine wave. Figure 3.24a is an accurate representation of the original signal, whereas Figure 3.24b is an aliased version whose frequency is much lower than that of the true signal. There is no way of detecting that a signal has become aliased just by looking at it and once aliasing has occurred there is no way of recovering the original signal, which is why all the emphasis is on prevention. There are also situations in which we do not have the luxury of being able to oversample the signal and we have to work close to the Nyquist limit. Under these conditions it is important that anti-aliasing filters are used to low-pass filter the signal before it is sampled.

Ignoring the fact that a digitized signal is not the same as the original can lead to various problems. For example, the sampled signals in Figures 3.23a and 3.23b contain all the information in the original sine wave. One way of finding the amplitude of the original sine wave is to measure its peak-to-peak height. If this is done on the sampled version of the sine wave in Figure 3.23b the result would be inaccurate, because the digitized signal is quite unlike the original. There are several ways to avoid the problem, and they are covered in detail in Chapter 5. Spectral analysis can be used or the signal can be resampled at a higher rate to make the sampled signal more like the original.

DIGITAL-TO-ANALOGUE CONVERSION

Quite often, a signal that has been sampled and stored in digital form needs to be converted back into analogue form. An example of this is compact disc players: the music is on the disc in digital form and has to be converted back to analogue form in order to listen to it. If the signal was sampled correctly, all the information in the original signal is present in the digitized version along with some extra high-frequency components. The original analogue signal is obtained by converting the digitized, sampled signal into an analogue representation and then removing the extra high-frequency components with a low-pass filter. A low-pass filter used in this manner is called a *post-sampling* filter or a *reconstruction* filter. It is very easy to determine what frequencies the filter should pass and what it should block because all the original signal components have frequencies below the Nyquist frequency (i.e. below half the sampling rate) and all the additional components have frequencies above the Nyquist frequency. The low-pass reconstruction filter should, therefore, have a corner frequency equal to the Nyquist frequency.

The digital signal is converted to analogue form with a digital-to-analogue converter (DAC). Using DACs is relatively easy compared to sampling and digitizing signals. One important point is that the analogue representation of the digital signal that a DAC produces has a 'staircase' pattern (Figure 2.9). The output of the DAC stays at the level of the last sample until the next sample is sent to it by the computer, resulting in the 'staircase' pattern. The sharp corners of the 'staircase' contain high-frequency components that are removed with the low-pass reconstruction filter.

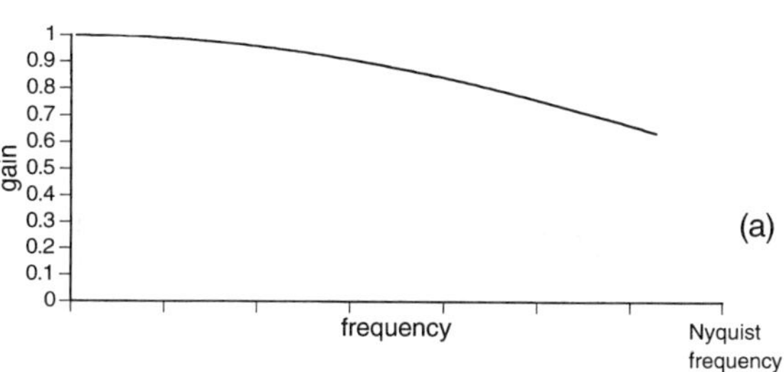

Figure 3.25 A DAC has an amplitude response that decreases with increasing frequency (a). Adding a filter with the opposite frequency response (b) will compensate for this.

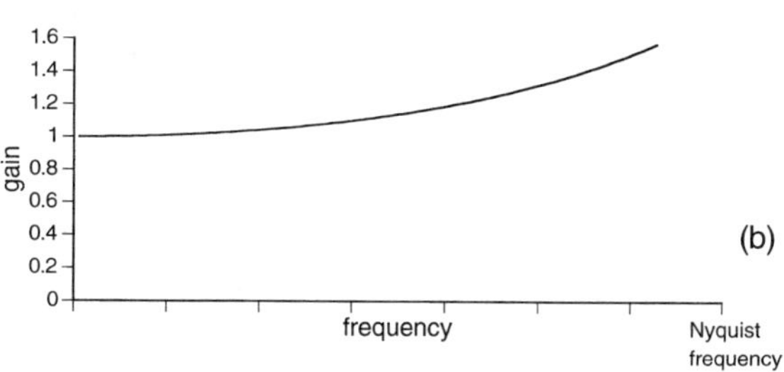

One less obvious problem with DACs is that they do not have a flat frequency response (Rabiner and Gold, pp. 301–2). The higher frequencies in the signal are reduced in amplitude. The relationship between the gain of the DAC and frequency is

$$\text{gain at frequency } f = \frac{\sin(\pi f\, T)}{\pi f}$$

where

$$T = \frac{1}{\text{sampling frequency}}$$

A graph of the gain against frequency is shown in Figure 3.25a. The reduction in high frequencies is worse when the frequency of the signal is close to the Nyquist frequency. One way to reduce the problem, therefore, is to sample at a higher rate. For a signal of a given frequency this will reduce the high frequencies less because the Nyquist frequency has moved further away from the signal frequency. If this is not possible and it is important that the signal amplitude is not altered then a common practice is to use a filter with the opposite frequency response (Figure 3.25b) along with the DAC so that the two effects cancel out. Some manufacturers make DAC boards with a built-in compensation filter.

chapter four

Connecting to the Real World with Transducers

A data acquisition system measures and records voltages. In order to measure useful quantities like pressure, temperature, force, etc., a *transducer*, a device that produces a voltage proportional to the quantity to be measured, is needed.

Most transducers do not produce a voltage in a range suitable for direct connection to a data acquisition system and some form of amplifier is needed. This may be a stand-alone module or may be incorporated right into the transducer housing. Usually the output voltage is proportional to the quantity being measured, but some sensors (thermistors, for example) are inherently nonlinear. Nonlinear sensors used to be a real pain to deal with but now the necessary linearization can easily be performed in software.

There is a vast range of transducers available for measuring almost any quantity, and only some of the common ones that have relatively simple interfaces will be covered here. More complex devices such as pH meters, ion-selective probes and glass microelectrodes are frequently used in life science work but all of these come complete with their own special amplifier and connecting the amplifier output to a data acquisition system is not difficult.

Only a few simple circuits are given here since the emphasis of this book is on collecting and analysing digital data rather than electronics. Chapter 15 in Horowitz and Hill goes into more details of the circuits used with different types of transducers. However, if you have the money in your budget it is nearly always better to buy transducer amplifiers and let someone else work out the details. What is important is knowing how to interpret manufacturer's specification sheets so that you can choose the best system for your application.

TEMPERATURE

The two classic types of temperature transducers used for temperature measurement are the *thermistor* and the *thermocouple*. Integrated circuit temperature sensors have also been developed that are convenient to use but have a narrower temperature range.

A thermistor is a piece of semiconductor material with two contacts on it. The material itself is a mixture of metal oxides such as manganese, nickel, cobalt, copper, iron, and titanium oxides which are fused into a small bead. The resistance of the bead changes with temperature. In the most common type of thermistor, the *negative-temperature-coefficient thermistor*, the resistance falls as the thermistor becomes hotter. This is contrary to what happens in metals, whose resistance increases with temperature. The fall in resistance with temperature is large, which makes measurements of small temperature differences easy. Thermistors are small, cheap, and rugged. On the down side, the change in resistance with temperature is nonlinear and any particular thermistor is only usable over a relatively small temperature range. Thermistors also do not work too well above about 300°C.

For the majority of applications the circuit in Figure 4.1 is fine for converting the change in resistance into a voltage. The thermistor is connected in series with a resistor, and a stable voltage applied across the pair. Most data acquisition systems have a stable reference voltage that can be used; if not, the +5 V supply can be used. With this configuration (Figure 4.1) the voltage recorded increases as the temperature increases.

The value of the resistor used requires some thought. Thermistors are specified by their resistance at 25°C (R_{25}). A value of 10–100 kΩ for R_{25} is good

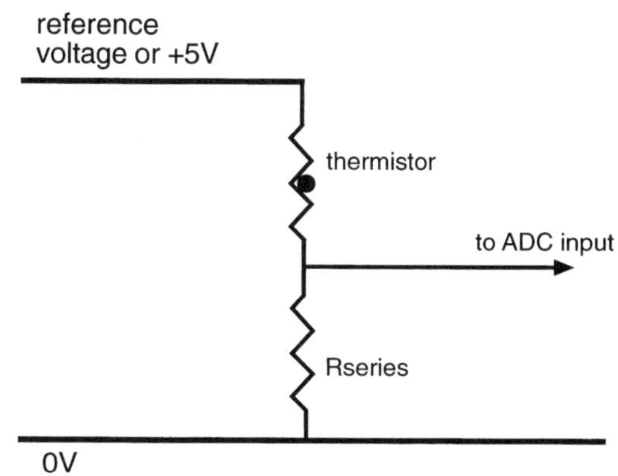

Figure 4.1 Simple circuit for interfacing a thermistor to a data acquisition system. With a careful choice of the value of the series resistor (R_{series}) the output voltage changes reasonably linearly with temperature. The reference voltage must be stable and free from noise.

for temperature measurement applications. The value of the series resistor, R_{series}, is calculated as a fraction of R_{25}. For measurements over the biological temperature range, about 10–40°C, a value for R_{series} equal to R_{25} gives reasonable linearity. For measurements at higher or lower temperature ranges the value of R_{series} should be adjusted. For optimum results the manufacturer's literature needs to be consulted because the value depends on the material from which the thermistor is made. As a guide, however, R_{series} should be about $3 \times R_{25}$ for measurements around 0°C and about $0.2 \times R_{25}$ for measurements around 70°C. Digi-Key has a good range of thermistors that are listed with their R_{25} values.

A value of 10–100 kΩ for R_{25} was stated as being good for temperature measurement applications. If the resistance is much lower than this the current flowing through the thermistor warms it up (*self-heating*) and produces errors. The smaller the physical size of the thermistor, the more likely this is to be a problem. If the resistance is too high then the circuit is likely to pick up noise.

Thermocouples are simply two wires of dissimilar metals connected together. Any two wires made of different metals (copper and iron, for example) and joined together will generate a small voltage that changes with temperature. In order to measure this voltage some sort of meter is connected across the two wires (Figure 4.2). The internal wire in the meter completes a circuit loop and thus the iron and copper wires meet each other twice, forming two thermocouples. One of these thermocouples is the main temperature sensing junction and the other is formed between the iron wire and the copper meter terminal. The voltages across the two junctions act against each other and thus the voltage measured by the meter is dependent not on the temperature of the sensing junction alone, but the temperature *difference* between the two junctions. The second junction that is formed at the meter terminals will be at room temperature. Even if the meter terminals are made from a metal other than copper, there will always be a second thermocouple junction formed somewhere.

The first thing to do when making a useful temperature measuring device based on thermocouples is to move the second junction to some place more convenient than the meter terminal, where its temperature can be controlled

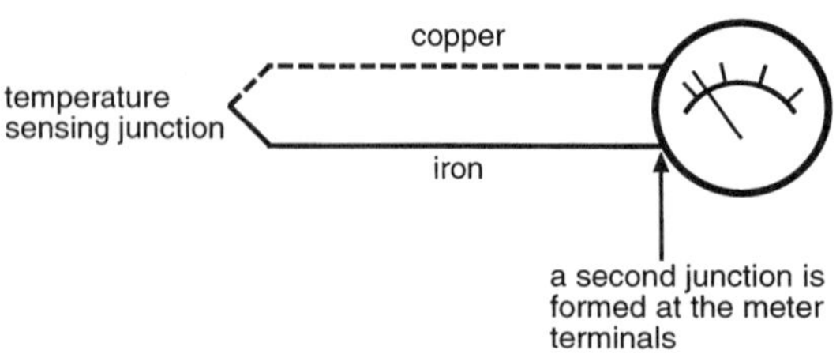

Figure 4.2 The basic circuit of a thermocouple-based temperature-measuring system. This layout is not very useful because the temperature of the second junction formed at the meter terminals is not controlled.

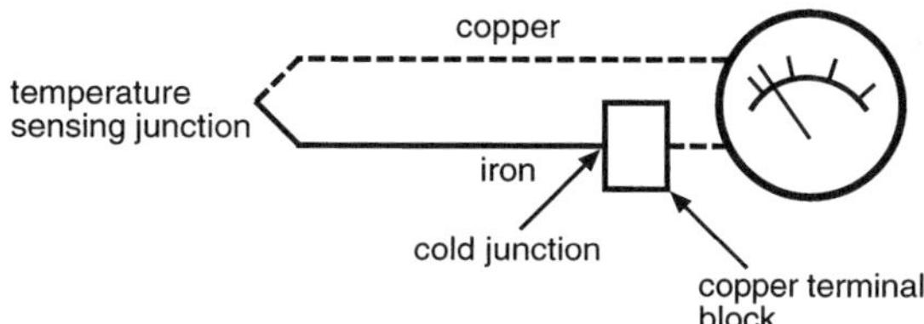

Figure 4.3 A practical thermocouple circuit. The second junction (the cold junction) is made at a defined place whose temperature can be controlled.

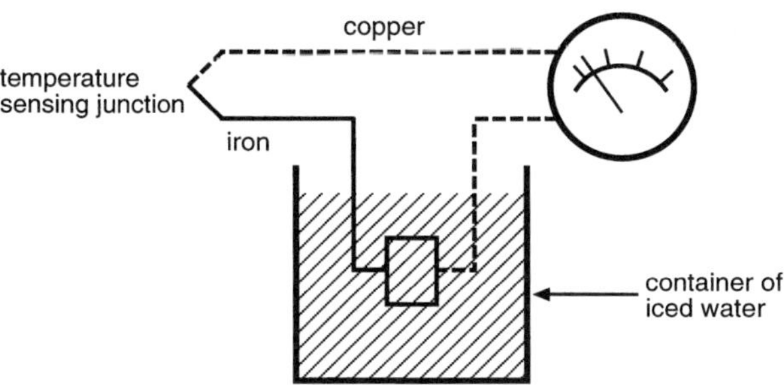

Figure 4.4 Cold junction compensation. The cold junction's temperature can be controlled directly by immersing it in a temperature-controlled bath. It is more convenient to measure the temperature of the cold junction with a second thermometer and correct the thermocouple reading.

(Figure 4.3). This second junction is known as the *cold junction*, even if it is actually hotter than the sensing junction. The cold junction is a block of metal made from the same alloy as one of the thermocouple wires. Now the voltage measured by the meter is dependent upon the temperature difference between the sensing junction and the cold junction. The type of wire in the meter circuit and its temperature is unimportant. Care must still be taken not to form extra junctions on the measuring side of the circuit and in practice this means using the same type of metal for each lead right up to the cold junction. Manufacturers make extension cables and connectors out of thermocouple alloys for this purpose.

The second task is to control the temperature of the cold junction. One way is simply to maintain it at a known temperature, for example by placing it in iced water (Figure 4.4). In this way the temperature of the sensing junction is referenced to 0°C. This gives accurate results but is not very convenient, and the usual way round the problem is to measure the temperature of the meter junction with a second thermometer and correct the thermocouple voltage. This is known as *cold junction compensation.*

Iron and copper are not particularly good metals from which to make thermocouples. Several alloys have been developed which give a predictable, almost linear voltage change with temperature over a wide temperature range. A good, general-purpose thermocouple is type 'K', which is made from the junction of chromel (90% nickel, 10% chromium) and alumel (96% nickel, 2% aluminium, 2% manganese) wires. Types 'J' (iron/constantan) and 'T' (copper/constantan) are older thermocouple pairs that are still in use – 'constantan' is an alloy made from 55% copper and 45% nickel. Type 'N' is a new alloy pair with some advantages over type K, but it is not so readily available. For very high temperatures (up to 1700°C) type 'R' thermocouples (100% platinum/87% platinum, 13% rhodium) are used.

The voltage produced by a thermocouple is small and an amplifier is needed. For example, a type-K thermocouple with a temperature difference of 25°C between the junctions generates 1.00 mV. Thermocouples are very widely used in industry and there is a large selection of probes and amplifiers available. Because each type of thermocouple generates a different voltage for the same temperature difference, thermocouple amplifiers are specific to one type of thermocouple. An exception to this is type T, which is close enough to type K over the range from 0 to 50°C that it can be substituted if necessary. If you want to make your own amplifiers, the AD595 chip from Analogue Devices is a complete type-K amplifier with cold junction compensation that costs about \$20. Thermocouple wire can be bought on reels from companies such as Physitemp. Making a probe is simplicity itself: just strip the insulation from the ends of the wires, wrap them together and solder the junction. Thermocouple wire is available in sizes down to 40 AWG and the latter is excellent for making sensors that can be passed through a hypodermic needle or catheter, although a microscope and steady hands are needed to strip and solder the wires.

Ordinary silicon diodes can also be used to measure temperature. The relationship between the current passing through a diode and the voltage across it is given by the diode equation:

$$I = I_0(e^{qV/kT} - 1)$$

where I is the current through the diode, V is the voltage across it, q is the charge on an electron, k is Boltzmann's constant, T is the absolute temperature, and I_0 is the leakage current of the diode. At room temperature, with a few mA of current passing through the diode, the voltage across it is about 600 mV. This voltage decreases by 2.1 mV/°C increase in temperature. Thus if we pass a constant current through a diode and measure the voltage across it, that voltage falls by 2.1 mV/°C. On its own, a diode is not very useful as a temperature sensor because of the large offset voltage (600 mV) and the fact that the actual voltage is dependent upon the current through the diode (I) and the leakage current (I_0). Furthermore,

I_0 varies from one diode to another because it depends upon how the diode is constructed.

However, with careful design and construction methods a diode can be used as the temperature-sensing element in a circuit. National Semiconductor produces a neat range of integrated temperature sensors that give an output voltage directly proportional to the temperature. The LM35 sensor has just three connections: ground, power and output. The output voltage is 10.0 mV/°C and is guaranteed accurate to ±0.5°C at room temperature (LM35A and LM35CA version: typical accuracy is ±0.25°C) which is better than most laboratory-grade digital thermometers. The sensor can be used over the temperature range from −40°C to +110°C ('C' version), although both positive and negative power supplies are needed to obtain the dual polarity output. But above +2°C the sensor only needs a single power supply of +5V to +30V, which makes it very useful in biological systems. The output voltage is 10.0 mV/°C so over the temperature range from +2°C to +100°C the voltage will go from 0.02 V to 1.00 V. This does not give good resolution if the sensor is hooked up to the input of a 12-bit ADC with an input range of ±5V, because the step size is $10/4,095 = 2.44$ mV and thus the resolution is 0.24°C. There are three ways to improve the resolution:

1. Change the range of the ADC to ±1 V. The resolution is now 0.05°C.
2. Use a 16 bit ADC. With a range of ±5 V the resolution is 0.015°C.
3. Add a fixed gain amplifier between the sensor and the ADC. An amplifier with a gain of 5.00 will use the full range of the ADC, bringing the resolution up to 0.05°C.

Other sensors in the LM35 family include the LM34, which has an output of 10.0 mV/°F. Both are available from Allied Electronics.

STRAIN

Strain is the fractional change in length of an object as it is stretched or compressed. On its own, a measurement of strain is not very useful except in mechanical engineering applications such as measuring the material properties of bone. But strain measurement is a key element in the operation of many types of transducer. To measure force, for example, the force can be applied to a bar or ring of metal and the amount that the metal deforms measured (Figure 4.5). For small deformations the amount of strain in the metal caused by the deformation can be made to vary linearly with the applied force and thus if the strain can be measured, so can the force.

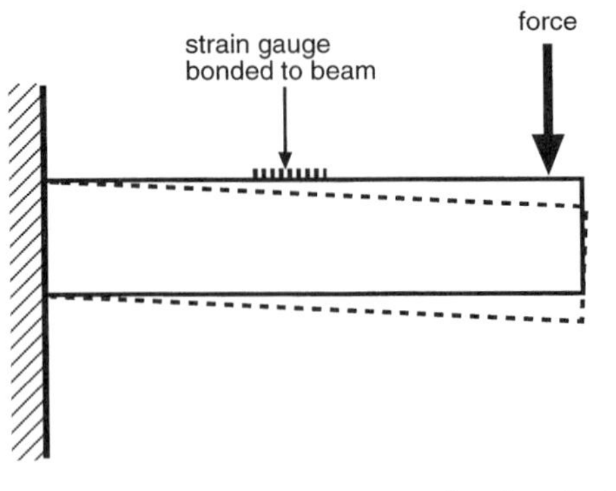

Figure 4.5 A simple force transducer. The metal beam bends elastically in response to the applied force, stretching the top surface of the beam and compressing the lower surface. The stretching and compression are measured with strain gauges.

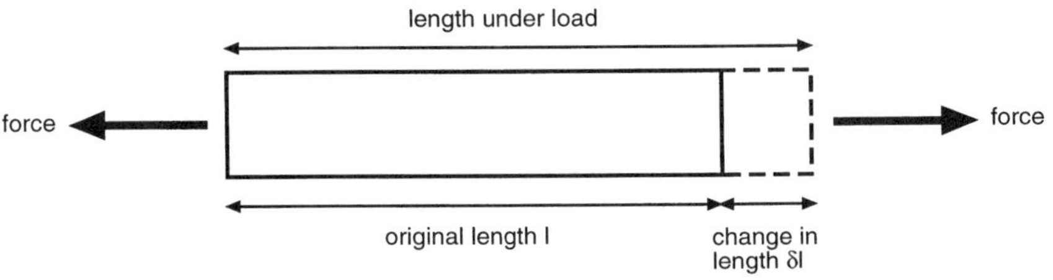

Figure 4.6 Definition of strain. When an object is stretched, it elongates by a small amount δl. Apparent strain is the increase in length (δl) divided by the original length (l). True strain is the increase in length (δl) divided by the actual length of the object ($l + \delta l$).

Strain (ε) is the change in length of an object divided by its original length (Figure 4.6) and thus (ε) has no units.[1] Practical strains in engineering materials such as metals are usually very small, much less than 1% and strain is often quoted in microstrain ($\mu\varepsilon$). A 1% change in length is therefore 10,000 $\mu\varepsilon$.

Strain is usually measured with a *strain gauge*. The most common type consists of a thin metal foil etched into a zigzag pattern (Figure 4.7) and protected on both sides by thin layers of plastic. The gauge is glued permanently to the object being deformed. As the object deforms, the gauge is stretched or compressed along with the object and this alters the electrical resistance of the gauge. The change in resistance is proportional to the strain.

The change in resistance of the strain gauge is tiny. A 120 Ω gauge stretched to a typical working strain of 1,000 $\mu\varepsilon$ changes resistance by only 0.24 Ω. The sensitivity of a strain gauge is expressed as the *gauge factor*. The gauge factor is defined as

$$\text{gauge factor} = \frac{\Delta R/R}{\varepsilon}$$

where R is the resting (unstrained) gauge resistance and ΔR is the change in resistance produced by a strain ε. A gauge factor of around 2 is typical for metal

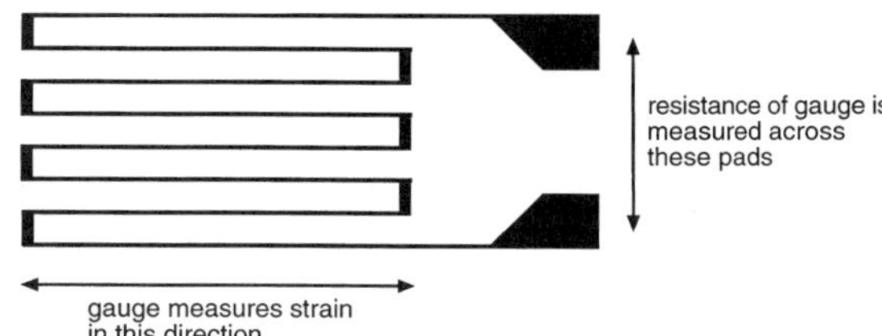

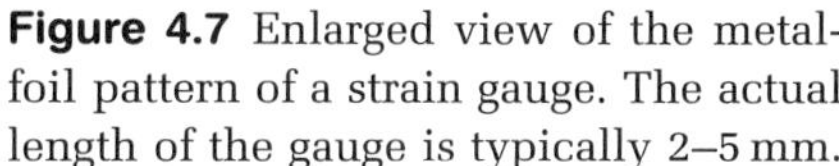

Figure 4.7 Enlarged view of the metal-foil pattern of a strain gauge. The actual length of the gauge is typically 2–5 mm.

foil strain gauges. Like strain, the gauge factor has no units. Metal foil strain gauges are available in standard resistances of 120 Ω, 350 Ω, and 1000 Ω.

Strain gauges can also be made out of silicon. These silicon strain gauges are much more sensitive than metal gauges, with gauge factors of several hundred, but are more susceptible to the effects of temperature. They can also be made very small indeed (1–2 mm long). Advances in silicon micromachining techniques have resulted in complete pressure and acceleration transducers made entirely from silicon: the strain gauge is embedded in the transducer itself.

The small change in resistance of metal foil strain gauges produces several sources of error due to temperature changes:

1. The resistance of the gauge changes with temperature.
2. The material to which the gauge is bonded changes dimension with temperature due to thermal expansion.
3. The wires connecting the gauge to its amplifier also change resistance with temperature.

Sources (1) and (2) can be arranged to cancel out to some extent by matching the thermal changes in resistance of the gauge with the coefficient of thermal expansion of the material to which the gauge is bonded. Strain gauge manufacturers sell gauges matched to metals commonly used in engineering such as mild steel, stainless steel and aluminium. Unfortunately for life scientists, strain gauges matched to materials such as bone and dental enamel are not available.

Trying to measure strain just by measuring the resistance of a gauge is not satisfactory because of the small change in resistance and the effects of temperature. These problems are solved elegantly by using a *Wheatstone bridge* circuit to measure the changes in resistance of the gauge.

The Wheatstone bridge is just four resistors connected in a square with a voltage applied across one diagonal (the *excitation* voltage) and a voltmeter across the other (Figure 4.8). The voltmeter will read 0 V if the bridge is *balanced*, that is if

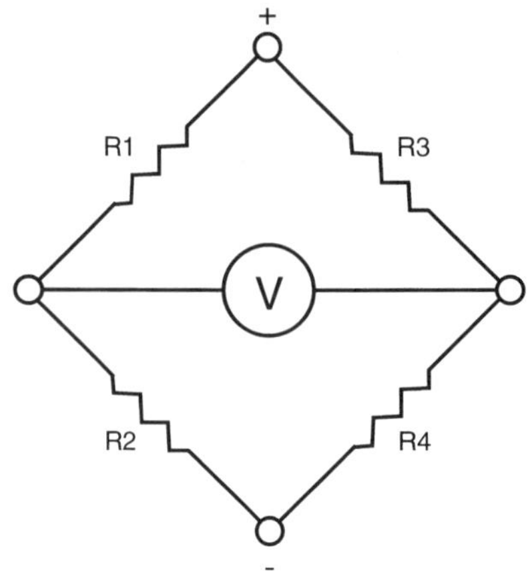

Figure 4.8 The Wheatstone bridge. A stable voltage is connected across the + and − terminals, and a voltmeter across the other two terminals. For small changes in resistance (ΔR) of one of the resistors the voltage measured is proportional to $\Delta R/R$.

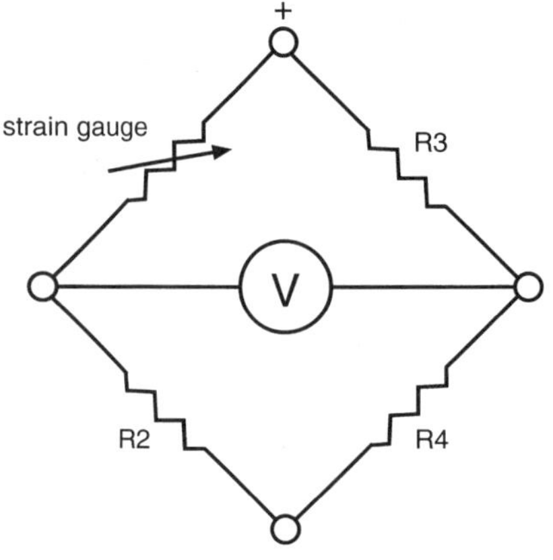

Figure 4.9 The quarter-bridge configuration for measuring strain with a single strain gauge. Here R_2 has the same resistance as the strain gauge. Both R_3 and R_4 have the same resistance.

$$\frac{R_1}{R_2} = \frac{R_3}{R_4}$$

For small changes (ΔR) in the resistance of any of the resistors (R_n) the voltage measured by the meter is proportional to $\Delta R/R_n$, making the bridge ideal for strain gauge measurements.

There are three configurations used to measure strain with a Wheatstone bridge. The *quarter bridge* configuration (Figure 4.9) uses one strain gauge and three 'dummy' resistors. The resistor R_2 must have the same resistance as the strain gauge; R_3 and R_4 just have to have the same value as each other. R_2, R_3, and R_4 need to be very high quality resistors whose change in resistance with temperature is predictable so that the ratio R_3/R_4 is constant. The quarter bridge

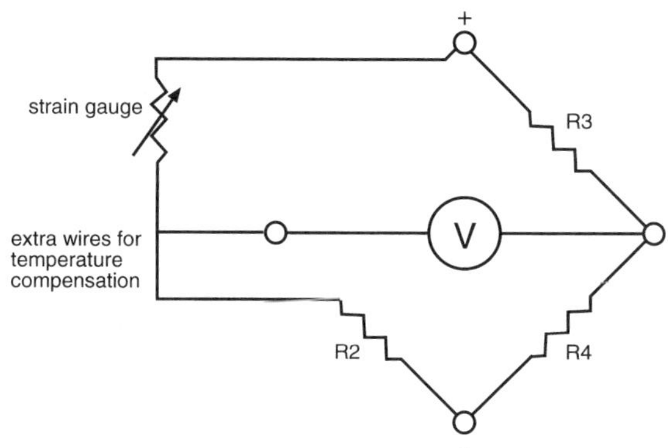

Figure 4.10 Compensating for changes in lead-wire resistance with temperature when using the quarter-bridge configuration. Extra lead wires are connected in series with R_2 so that any change in lead-wire resistance affects both sections of the bridge.

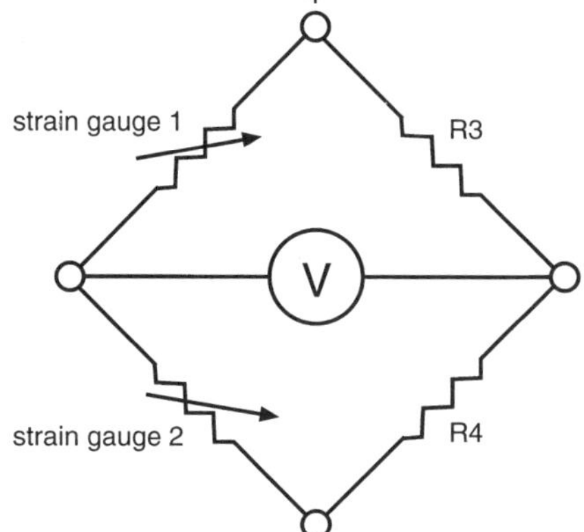

Figure 4.11 The half-bridge configuration for measuring strain with two strain gauges. The gauges are arranged so that one is stretched and the other compressed when the object is loaded. R_3 and R_4 have the same resistance.

arrangement is sensitive to changes in temperature of the connecting wires. Also, it is often difficult to find a resistor for R_2 that has the same change in resistance with temperature as the strain gauge, and even if this is possible the gauge and R_2 are usually not next to each other and may well be at different temperatures.

The problem of changes in temperature of the connecting leads can be overcome with an extra lead wire (Figure 4.10). Now any thermal changes in lead-wire resistance are cancelled out, as long as all the wires are at the same temperature and made of the same material. This is relatively easy to arrange by using four-core cable.

The *half-bridge* configuration (Figure 4.11) uses two gauges and two stable dummy resistors. The two gauges are mounted close together and are identical, so most of the temperature effects are eliminated. Furthermore, it is often easy to arrange things so that one gauge is stretched and one compressed (by mounting

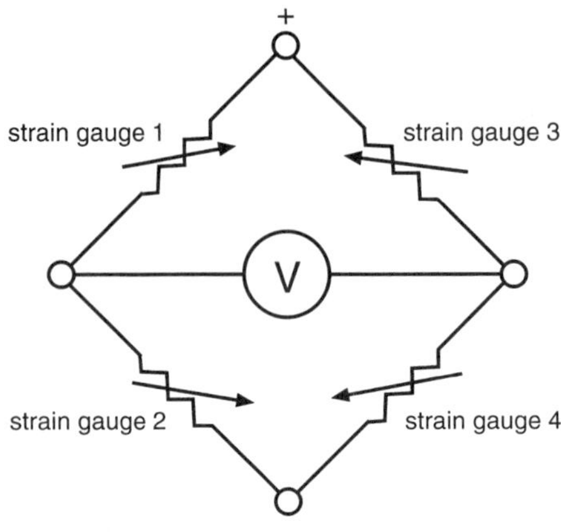

Figure 4.12 The full-bridge configuration for measuring strain with four strain gauges, two in compression and two in tension.

one on each side of a piece of metal). This doubles the sensitivity of the strain measurements.

The *full bridge* (Figure 4.12) uses four gauges. These are often mounted in pairs so that two are compressed and two stretched. A full bridge is not used that much for strain measurements, but is very widely used inside transducers. It is so common that many manufacturers make a general-purpose '*DC bridge amplifier*'. This is nothing more than a stable DC power supply and a sensitive amplifier/voltmeter inside one box. The power supply, whose voltage can often be varied, is used to excite the bridge. The amplifier/voltmeter measures the output of the bridge and displays it in a suitable format. The excitation voltage is variable because there is a trade-off: as the excitation voltage increases, the measurements are less susceptible to noise but the gauges start to warm up and self-heating becomes a problem. Typical values for bridge excitation voltages are 2–5 V. The amplifier/voltmeter usually has a gain that can be varied by the user, plus an offset control that is used to zero the output. Other bells and whistles that are often added are low- and high-pass filters that can be switched in to reduce noise.

FORCE

Force transducers come in two main types. The commonest is a *load cell*. This is nothing more than a lump of metal with strain gauges attached to it. Usually four gauges are used in a full-bridge configuration. By careful choice of the shape of the piece of metal the change in strain is linear with the applied force, which can range from grammes to tonnes. Load cells come in a wide variety of shapes: round and S-shaped cells for measuring compressive forces, beams (bars of metal that bend with the applied force), even bolts with strain gauges built right inside them so that the tightening force can be measured. Load cells are widely used in weighing applications. Because they use strain gauges as the measuring device, load cells use standard strain gauge amplifiers to give an output voltage proportional to the applied force.

The other type of force transducer is based on the piezoelectric effect. Many materials develop a charge when subjected to mechanical strain, the piezoelectric effect. The reverse is also true: when a voltage is applied, the material bends. Piezoelectric buzzers and alarms are very common in battery-powered equipment because they are small, cheap, and quite efficient. For example, the alarms in digital watches and smoke detectors are all piezoelectric.

The charge generated by a piezoelectric material when it is deformed is proportional to the strain and the effect can be used to make force transducers. The piezoelectric element is a ceramic, quartz or plastic material with a metallic electrode on each side. The element is then enclosed in a suitable metal case. Piezoelectric force transducers present some interesting problems for amplifier design. It is the charge generated by the force that is important, and thus *charge amplifiers* are needed to get a voltage output that is proportional to force. The charge on the transducer element slowly leaks away with time and piezoelectric force transducers cannot measure constant forces, only changing ones. On the other hand, they can measure extremely rapidly changing forces because of the small mass of the piezoelectric element. A typical frequency range for a piezoelectric force transducer is 2 Hz to 10 kHz. They find extensive use for measuring rapidly changing forces such as those occurring during vehicle crash testing and sporting activities.

An interesting biological application of load cells is the force plate. This consists of a stiff plate with a set of load cells at each corner. The load cells in each set are arranged to measure the force in all three axes (forward–back, left–right and up–down). By combining data from all the sets of load cells the total force on the plate along any axis, plus the three turning forces around the axes, can be resolved. Force plates are used extensively in biomechanics research. For example, the resilience of different types of footwear can be analysed by measuring the forces generated when a person runs across the plate. Another application is the assessment of treatment regimens for orthopaedic conditions. The pattern of forces generated by the foot of a person or animal as they walk across the plate will change if the limb hurts, and thus the effect of different treatment regimens can be measured objectively. Kistler (*www.kistler.com*) is a major manufacturer of force plates for both research and clinical use.

ACCELERATION

If a mass is attached to a force transducer it can measure acceleration using the relationship

$$f = ma$$

where f (N) is the force required to accelerate a mass m (kg) by acceleration a (m s^{-2}). The mass m is often called a *seismic mass* and is just a blob of metal, silicon or whatever is handy. Accelerometers are widely used to detect movement and vibration. Because they are based on force transducers they also use either strain gauges or piezoelectric elements as the basic sensor. Strain gauge devices will measure constant accelerations (such as that due to gravity) whereas piezoelectric accelerometers only measure changing acceleration, but have a better high-frequency response.

Force and acceleration are vector quantities, that is, they have both magnitude and direction. In three-dimensional space, any force or acceleration can be reduced to three forces or accelerations at right angles to each other. Force and acceleration transducers are available that measure one, two, or all three of these components (the latter are referred to as *biaxial* and *triaxial* transducers, respectively). The direction in which they measure the forces or acceleration is marked on the outside of the transducer and it is important when mounting the transducers to make sure that they are lined up correctly. A transducer should only respond to a force or acceleration along its measurement axis. The *transverse sensitivity* is a measure of how well it rejects force or acceleration in other directions.

A range of low-cost accelerometers are made by Motorola (XMMAS series), ICSensors (a division of EGG) and Analog Devices (ADXL series). The working parts of the transducer – seismic mass, beam, strain gauges, and end stops – are all etched from a single piece of silicon. This greatly reduces the cost and these accelerometers are ideal for situations in which the gauge might be lost or damaged.

Another advantage is that the strain gauge amplifier can also be fabricated on silicon. The result is a single package whose output is a filtered and amplified acceleration signal all ready to hook up to a data acquisition system. For example, the ADXL150 from Analog Devices (available from Allied Electronics) contains an accelerometer, amplifier, low-pass filter, and self-test function in one package. The range is ± 50 g with less than 0.2% nonlinearity and 2% transverse sensitivity. With an output sensitivity of 38 mV g^{-1} and a single $+5$ V supply the output range is 0.6 V–4.4 V (-50 g to $+50$ g) which perfectly matches an ADC with a $+5$ V unipolar input. Analog Devices also makes two-axis accelerometers in the same range (the ADXL250, for example).

An interesting application of accelerometers in the behavioural sciences is measuring the 'startle response' of small rodents. Rats and mice exhibit a strong startle response in response to a sudden noise, which might indicate an approaching predator. The startle response is similar to that in humans who 'jump' when surprised by a sudden noisc. The response can be measured by placing a rat or mouse on a platform that has an accelerometer attached to the underside. When the animal 'jumps' in response to a noise the sudden change in body position changes the ground reaction force, and this changing force causes the plate to move slightly. The plate movement is recorded by the accelerometer. The duration of the acceleration is about 100 ms in rats and 60 ms in mice and the signal is not repetitive so a reasonably fast sampling rate of about 1 kHz is used to record the acceleration. The acceleration can be quantified by its maximum amplitude, duration, area under the acceleration–time curve, and so forth. Anxiolytic drugs and other drugs with CNS effects will change the startle response and the accelerometer plate can be used to quantify the effect.

DISPLACEMENT AND ROTATION

Displacement transducers measure how far, and in which direction, something has moved. The most common displacement transducers have a small plunger that moves in and out of an outer barrel (Figure 4.13). Typically, the plunger is attached to the mobile object under investigation and the barrel is fixed to the bench. The transducer output is proportional to the distance that the plunger moves relative to the barrel. The plunger can be free-moving, in which case it has to be attached to the mobile object, or spring-loaded, in which case it only has to touch the object. Displacement transducers are available that can measure total

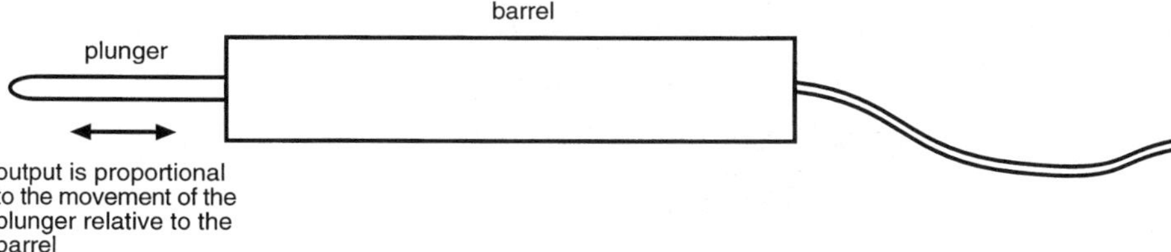

Figure 4.13 A typical displacement transducer. The plunger moves freely in and out of the barrel, and the output of the transducer is proportional to the movement of the plunger relative to the barrel. The plunger may be free-moving or have a spring to return it to its fully extended position.

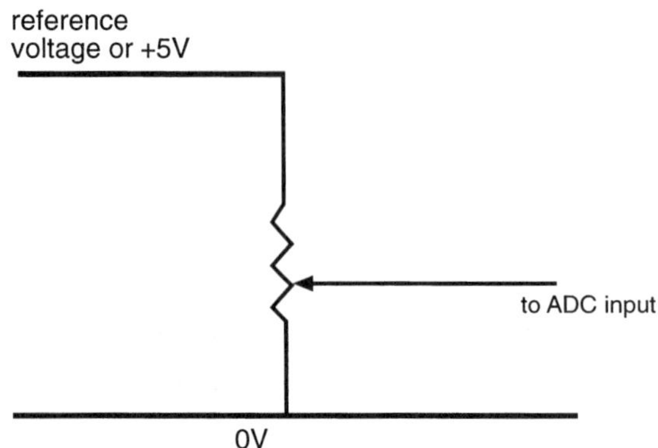

Figure 4.14 Circuit for interfacing resistive-type displacement transducers to a data acquisition unit. A stable voltage is applied across the ends of the track and the output is taken from the sliding contact.

displacements in the range 1–940 mm, with resolution down to the micron range. Displacement transducers are widely used in industry and aviation and there are a number of manufacturers (e.g. Penny & Giles, *www.penny-giles-control.co.uk*) whose products are widely distributed.

Two main principles are used to measure the movement of the plunger inside the barrel. The simplest type is resistive. Inside the barrel is a track made from carbon-based material which has a constant resistance per unit length. A sliding contact on the plunger moves along this resistive track so that the electrical resistance from one end of the track to the sliding contact is proportional to the distance between the end of the track and the contact. Usually you have access to both ends of the track and the sliding contact, and a potential divider circuit is used to generate a voltage proportional to the distance moved (Figure 4.14). These potentiometric displacement transducers are by far the easiest to interface to an ADC because no amplifier or external components are needed. The range of the ADC can be matched perfectly by making the voltage applied to the ends of the track the same as the range of the ADC.

The second type of displacement transducer is the *variable-reluctance* device. Just like the variable-reluctance pressure transducer (see below), these transducers have a set of fixed coils fed with an AC signal and a sense coil wound on the inside of the barrel. The plunger is made from a magnetic material and as it moves in and out of the barrel the signal generated in the sense coil varies in strength. The amplitude of the signal is proportional to the displacement. Like variable-reluctance pressure transducers (see the next section), a specialized AC bridge amplifier is needed. Variable-reluctance displacement transducers are very commonly referred to as LVDTs (linear variable-displacement transformers). The term LVDT is also frequently applied to resistive-type displacement transducers as well. Variable-reluctance transducers have very high resolution and a long life because there is no track or sliding contact to wear out.

A third type of displacement transducer uses an *optical encoder*. A strip of metal has alternating reflective and nonreflective bars on it about 0.04–0.2 mm apart. An optical system bounces light off the strip and measures how much is reflected. Resolution is typically 1/4 of the width of the bars, and by using more than one optical detector the direction of travel is known. The output of the optical encoder is a series of pulses which have to be processed further to give a distance. Another problem is that there is no absolute measure of distance, only relative. To measure absolute distances the transducer has to be zeroed at some convenient location and all measurements are taken relative to this point. Optical displacement transducers are not that useful in life-science work but are widely used in industry for controlling machinery. One great advantage of optical encoders is that the reflective strip can be stuck onto some part of the machine such as one of the carriage rails on a lathe. The optical encoder rides on the lathe carriage. The reflective strip can be as long as desired, which may be several metres.

Rotary motion is very similar to linear motion and transducers based on resistive and variable-reluctance principles are available. Again, resistive transducers are the easiest to interface to an ADC. These transducers are just precision potentiometers. They may span a single turn or many turns, five- and ten-turn designs being common. Single turn rotary motion sensors are also called angle sensors and have a typical range of 300°. They can be resistive or variable-reluctance. Like the linear variable-reluctance transducer, an AC exciter/amplifier is needed for variable-reluctance angle sensors but linearity and resolution are excellent. If you take an angle sensor and hang a weight on the shaft to act as pendulum, you have an inclinometer that will measure tilt angles. Sensorex makes a range of inclinometers with ranges from $\pm 1°$ to $\pm 90°$ that are complete units with integrated amplifiers.

Optical encoders are also used for rotary transducers. Like the linear optical encoders, the output of the sensor is a series of pulses. Typical resolution is 50–1000 counts/revolution.[1] Optical rotary encoders can be used to determine angle but by far their biggest use is to measure shaft rotation speed. In the simplest application the frequency of the pulses is measured and converted to a rotation rate, in rpm, for example. In more complex applications the number of pulses (after some suitable reset) is counted and thus the number of shaft rotations is known.

Rotary transducers can be modified to measure displacement to produce a device that is very useful in life science work. The *cable extension transducer* consists of a thin steel cable wound around a drum. The drum is spring-loaded so that it automatically reels the cable back in, just like a steel tape measure. A rotary transducer is attached to the shaft of the drum. Because the diameter of the drum is constant, the amount of cable that is pulled off the drum is known accurately. The end result is a displacement transducer that is easy to attach to a person and which will extend to considerable distances, making it very useful in kinematics and sports-medicine research. A huge advantage over LVDTs is that the angle of the cable relative to the transducer can change. For example, the end of the cable can be attached to the seat of a rowing scull. The position (and therefore the velocity and acceleration) of the seat can then be recorded with good accuracy. At the same time, another transducer attached to the shoulder will monitor upper-body position: differences between the two signals reflect changes in upper-body angle. Cable extension transducers are made by several companies including Space Age Control Inc. (*www.spaceagecontrol.com*).

Finally, specialized angle sensors are available to specifically measure joint angle. Joint flexion/extension is measured with a goniometer and joint rotation is measured with a torsiometer. These sensors are small and light enough that they can be taped to the wrist or finger without interfering with normal movements. In rehabilitation work the range of joint motion can be accurately and objectively measured. Joint extension after surgery can be used to assess the adequacy of post-operative analgesia in humans and animals. Wrist and finger motion can be monitored to investigate repetitive motion disorders. One manufacturer of goniometers and torsiometers is Biometrics Ltd. (*www.biometricsltd.com*).

PRESSURE

Like temperature, pressure is a quantity that frequently needs to be measured in industry and a large variety of pressure transducers is available, most of which use a similar operating principle. Inside the transducer is a thin diaphragm which is usually metallic. The pressure to be measured is applied to one side of the diaphragm and a reference pressure (either atmospheric pressure or a vacuum) is applied to the other. The diaphragm bends proportionally to the applied pressure and this bending is detected with strain gauges bonded to the diaphragm. Any DC bridge amplifier can then be used to give an output proportional to the pressure. The wide range of pressures that can be measured is accommodated by changing the thickness of the diaphragm.

Pressure transducers are available in various flavours. They all inherently measure a difference in pressure between the two sides of the diaphragm. The pressure on the reference side can be atmospheric, in which case the output is relative to atmospheric pressure: *gauge pressure*. If the reference side is a vacuum the output is *absolute pressure*. And thirdly, pressure can be applied to both sides of the diaphragm in which case the output is proportional to the difference between the two pressures: *differential pressure*. Differential pressure transducers are commonly used to measure gas flow in a tube, by measuring the pressure drop across some restriction in the tube.

There is a large range of strain gauge pressure transducers available for industrial use, but most of these are not suitable for life-science work. Because they have to cope with an industrial environment, they are enclosed in a solid metal case and are heavy and bulky. They also tend to be rather insensitive. One exception is Gaeltech (*www.gaeltech.com*) who make metal diaphragm, strain-gauge-based pressure transducers specifically for the medical and life science market.

Invasive blood pressure measurement is an important part of anaesthesia and critical care, and there is a large market for disposable pressure transducers. These are good for pressure measurement over the range 0–300 mmHg. They use silicon strain gauges to detect the bending of a moulded plastic diaphragm. The transducers are normally plugged into a patient monitor that displays the pressure waveform along with systolic, diastolic, and mean pressures. However, they can easily be adapted for other pressure measurements using a standard DC bridge amplifier.

Motorola, Sensym (Milpitas, California, USA) and Fujikura (*www.fujikura. com*) make pressure transducer elements entirely from silicon: the diaphragm is a thin sheet of silicon with the strain gauges formed right inside it. The basic pressure sensing element has the usual properties of silicon strain gauges: high sensitivity (large gauge factor) but marked changes with temperature. The temperature problems are corrected in the 'compensated' versions with a temperature sensor built into the element. Even better, a bridge amplifier circuit can also be formed in the silicon using standard integrated circuit techniques so that the final package contains both the transducer and bridge amplifier. These transducers are simple to use: apply a stable 5 V supply and the output is a voltage between 0 and 5 V that is proportional to pressure. They are available in absolute, gauge, and differential versions. Silicon pressure transducers are small (especially the ones made by Fujikura), lightweight, sensitive, inexpensive, and have a fast response, which makes them ideal for life-science work.

A different type of transducer is made by Validyne and Celesco. They still use a diaphragm that bends with the applied pressure, but the bending is not detected with strain gauges. Instead, two coils of wire are mounted close to the diaphragm and fed with an AC signal of about 5 kHz. A third coil wound with the other two picks up the resultant magnetic field. The AC signals fed to the two main coils are 180° out of phase with each other, so under normal conditions (with no pressure applied) they cancel each other out and there is no voltage induced in the pick-up coil. If the diaphragm bends slightly, the signal from one of the main coils is received more strongly than the other by the pick-up coil and thus an AC voltage whose amplitude is proportional to the applied pressure is generated in the pick-up coil. This is amplified and converted to a DC voltage proportional to the applied pressure by an *AC bridge amplifier*. Transducers which use this principle are called *variable-reluctance transducers*. For a long time their performance was superior to strain gauge type transducers, particularly for low pressures, but the silicon pressure transducers are as good and a lot less expensive. Variable-reluctance transducers tend to be heavy and must be used with an AC bridge amplifier, which limits their usefulness.

LIGHT

Transducers that detect light intensity are used both as light meters (spectrophotometry, exposure meters, etc.) and as yes/no detectors (position sensors, optical data links, etc.). For accurate measurement of light level a calibrated sensor is needed, and the spectral range of the light must be considered and restricted with filters if necessary. But in many cases a simpler detector that produces an output proportional to ambient light level is all that is needed.

The simplest device to use is the cadmium sulphide/cadmium selenide *photoconductive cell*, also called a *light-dependent resistor*. This device has a resistance in the dark of about $1\,M\Omega$, falling to a few thousand ohms in bright light. This large change in resistance makes it easy to interface to a data acquisition system (Figure 4.15). The drawbacks of the device are its slow (200–300 ms) and nonlinear response. It is best used for light levels that change slowly and a common

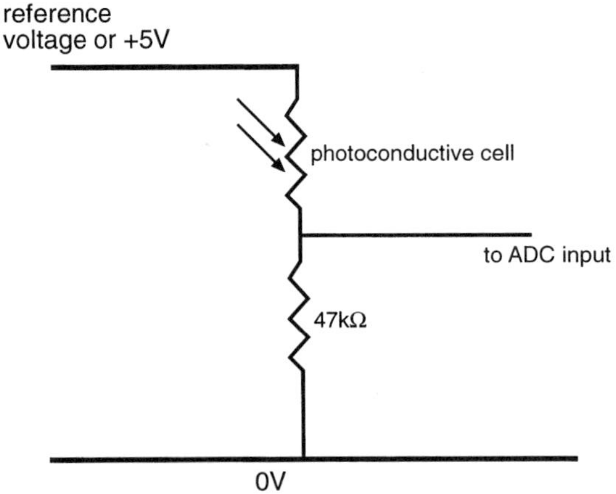

Figure 4.15 Simple light detector using a photoconductive cell (light-dependent resistor). The output is nonlinear and has a relatively slow response. In this configuration the voltage on the ADC input increases as the light level increases. To reverse this, swap the resistor and the photoconductive cell.

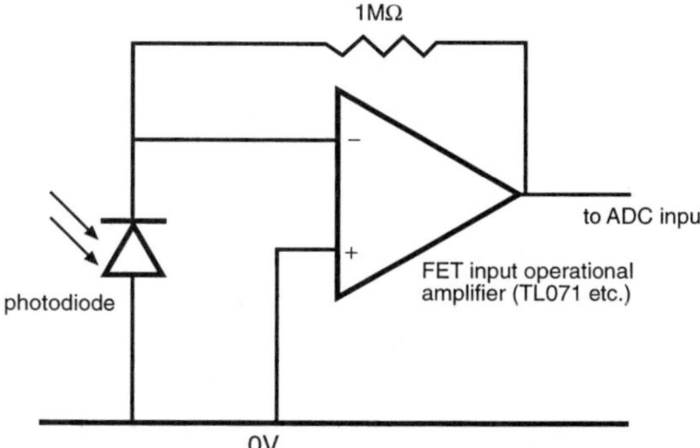

Figure 4.16 A photodiode and operational amplifier make a fast, linear light meter.

application is in lights that turn on automatically at dusk. For example, Vactec (a division of EGG) sell the VT43N3 specifically for automatic street lights. The VT935G is designed for use in night lights, toys and cameras. Both are very inexpensive and are available from Allied Electronics. Allied also carry a complete range of photoconductive cells made by Silonex.

Photodiodes and phototransistors are widely used to detect and measure light levels. They are small, cheap and have a fast response. A photodiode generates a small current that is proportional to the light level because each photon that is absorbed by the diode generates an electron–hole pair. Since data acquisition systems measure voltages, a current-to-voltage converter is needed. One operational amplifier and one resistor is all that is needed (Figure 4.16) to make a highly linear light detector. The gain of the circuit is set by the resistor. The higher the value of the resistor, the more sensitive the circuit is to low light levels and some designs call for resistor values of $100\,\text{M}\Omega$. Several problems creep in when such high value resistors are used: not only are they difficult to obtain but the resistances of the circuit board, wires, and so forth, become significant. The circuit also becomes very sensitive to electrical noise, and will saturate in bright light. Horowitz and Hill (pp. 996–7) go into the circuit design a little more, and for detailed analysis the Burr-Brown application notes AB-075, AB-077, and AB-094 are useful (*www.burr-brown.com*).

The current generated by photodiodes is small and it is important that the leakage current of the operational amplifier (op amp) does not swamp it. An op amp with FET inputs is necessary to obtain the low leakage current. For general-purpose use the TL071 op amp from Texas Instruments is cheap and readily available from Digi-Key, Allied Electronics, and many others. It is also available in a dual (TL072) and quad (TL074) version. The pin connections are shown in Table 4.1.

TABLE 4.1 **PIN CONNECTIONS FOR THE TL071 RANGE OF OP AMPS. TL071 IS A SINGLE DEVICE, TL072 A DUAL, AND TL074 A QUAD. FOR MULTIPLE OP AMPS THE PINS ARE GIVEN IN ORDER, I.E. THE TL072 AMPLIFIER 1 HAS + VE INPUT ON PIN 3, − VE INPUT ON PIN 2 AND OUTPUT ON PIN 1**

Device	Positive input	Negative input	Output	Positive supply	Negative supply
TL071	3	2	6	7	4
TL072	3,5	2,6	1,7	8	4
TL074	3,5,10,12	2,6,9,13	1,7,8,14	4	11

Silicon photodiodes have a peak response in the infrared region around 850–900 nm. This is exploited in infrared remote controls, sensors, and wireless computer links. In these applications you want to exclude visible light and it is common for photodiodes to be packaged in plastic that is transparent only to infrared light. Digi-Key has a wide range of infrared photodiodes and matched infrared LEDs. If you really want to measure visible light then make sure the diode has a clear window on it. Silonex has a range of suitable photodiodes which are distributed by Allied Electronics. You still have the problem that silicon diodes are more sensitive to infrared light than visible light. If it is important that the light sensor's response is similar to the mammalian eye then a blue-green filter can be used in front of the diode. An alternative approach to using a separate photodiode and operational amplifier is the OPT301 from Burr–Brown (available through Digi-Key) which combines a photodiode and high performance op amp in one package.

Phototransistors combine a photodiode with a transistor so that the current generated by light falling on the transistor is amplified by the same device. Phototransistors are very easy to interface to data acquisition systems using just one resistor (Figure 4.17). The drawback is that the output is nonlinear and easily 'saturated' – the circuit responds to a rather narrow range of light levels, and light intensities outside this range produce an output that is fixed high or low. This is fine if all you want to do is detect whether light is present or not, for example, to record when a light beam is broken by an animal's movement. Phototransistors are also much slower than photodiodes with a maximum frequency response of a few kHz.

If you need to record light levels and/or need a fast response, use a photodiode. If you need a yes/no detector, a phototransistor is simpler to interface to the data acquisition system. Again, Digi-Key and Allied have good ranges of phototransistors. LED/phototransistor pairs are often used as sensors to detect whether an

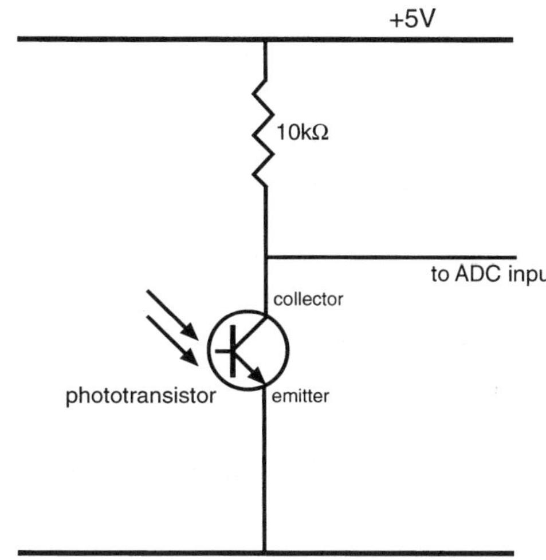

Figure 4.17 Phototransistors make cheap light detectors that are easy to interface. The output is nonlinear and has a slower response than the photodiode system (Figure 4.16).

object is breaking the infrared beam or not. Even with infrared filters, ambient light can be a major interference problem. The way to solve it for single transmitter–receiver systems is to modulate the LEDs so that the light flashes on and off at a few kHz. The phototransistor is hooked up to a tuned amplifier that only responds to the modulated signal. Such an approach is essential when long beams are needed.

LED/phototransistor beam systems can be used to detect when an animal moves between them and form the basis of an activity monitoring system. The infrared beam is invisible and does not disturb the animal. In the laboratory the separation of the LED and phototransistor is not far, often the width of a rodent cage, and simple systems can be used. Use an LED with a narrow beam angle and align it carefully with the phototransistor, which should have an infrared filter. The ACTI+ from Viewpoint (Lyon, France) is a commercial system that uses an infrared beam to monitor rodent activity. Another system is made by AccuScan Instruments Inc. (*www.accuscan-usa.com*). The AccuScan system uses arrays of sixteen sensors, 2.5 cm apart, along each side of a square cage. One pair of sensor arrays is placed on the front and back of the cage and a second pair is placed on the two sides. The crossed beams divide the cage into a grid with each cell being 2.5 cm × 2.5 cm. A third pair of sensor arrays is mounted higher up the cage. The height is such that the beams are only broken when the animal stands up, giving a third dimension of movement that can be recorded.

Three sets of sensor arrays contain 48 individual beams, all of which have to be continuously monitored to see if they are broken or not. Wiring each sensor up to an individual digital i/o line is possible but inconvenient, particularly because standard A/D cards do not have forty-eight digital lines. To get around this problem the sensors in each array are *multiplexed*. The sixteen LEDs in each array are turned on sequentially and only one LED is on at any one time. The speed at which the LEDs are sequenced is fast enough that the animal does not move very far in the time that it takes to turn all sixteen LEDs on and off. The reduction in the number of wires going to each sensor array is substantial. Without multiplexing you need one wire per LED plus a common wire, seventeen in all. With multiplexing you only need four wires to code for all sixteen sensors ($2^4 = 16$) plus a common wire, five in all. Another advantage is that the LED and phototransistors do not have to be aligned very precisely. In fact, you can have an array of phototransistors that are all connected together so that they respond to infrared light anywhere along the length of the array. Because you know which LED is on, the signal received from the phototransistors must come from that LED and all you need to do is know whether the beam is blocked or not. The price that you pay is increased complexity of the circuitry because the multiplexing needs to be performed in hardware. Also, you still have to get the information on which beams are broken into the computer. The digital i/o port can be used for this but the AccuScan system encodes the data and sends it down the serial port to the host computer. This greatly increases the amount of circuitry needed but it does mean that an A/D card is not needed on the host computer.

ISOLATION AMPLIFIERS

Measurements of biopotentials (ECG, EMG, EEG), tissue impedance, and so forth pose special risks because the subject or animal is connected via a low-resistance pathway (the electrode) to a piece of electrical equipment. If a fault should develop in the latter there is a real risk of electrocution. All medical equipment must conform to strict electrical safety standards which protect the patient in the event of a fault or other unusual situations such as electrical defibrillation. One of the basic principles is that all circuitry that is connected to the patient must be insulated from the mains supply and the equipment casing. The insulation has to have very low leakage (μA) and must be able to withstand several thousand volts. The way this is achieved is by the use of isolation amplifiers. Inside each amplifier module is an insulating barrier across which the signal information is transmitted. Common ways of doing this are to use transformers (Analog Devices AD295), convert the signal to light (opto-isolators), or use the small capacitance of the insulating barrier (Burr–Brown ISO102). The patient side of the equipment consists of the electrodes and an amplifier to boost the signal to 1 V or 2 V. The isolation amplifier passes this signal across the barrier to the mains side of the system, where it is amplified and displayed using conventional circuitry. Apart from the insulation requirements, the other big problem is powering the circuit on the patient side because the power supply must also be insulated. Batteries are one popular option. Isolated power supplies are another, but their output is limited to a few mA in some cases.

In many research settings it is desirable to collect some bioelectrical signal from an animal or human subject. It is very tempting to use a standard laboratory amplifier and data acquisition system, rather than paying the premium for an isolated system. But the laboratory amplifiers will not protect the subject in the event of equipment failure and the results could be tragic. Some data acquisition manufacturers (e.g. ADInstruments) sell isolated amplifiers to go with their systems, and outdated medical ECGs are often available from hospitals at low cost. Another option is to use battery-powered amplifiers and a laptop computer to collect the data. But do not make the mistake of running the laptop from an external power supply when collecting data – use the internal batteries!

NONLINEAR SENSORS AND LINEARIZATION

A linear transducer is one for which a plot of the output voltage against the quantity being measured is a straight line (Figure 1.1). Linear transducers are easy to calibrate using a two-point calibration procedure (Chapter 1).

Nonlinear transducers have an output that does not vary linearly with the quantity being measured, that is, a plot of the output voltage against the quantity being measured is curved (Figure 4.18). Nonlinear transducers used to be a real problem to deal with before personal computers were readily available and great efforts were made to linearize the output.

The general way to deal with nonlinear transducers is to try to find a mathematical relationship between the input and the output. Once this is known, it is a relatively simple matter to get the computer to use the relationship to correct for the nonlinearity. Suppose the relationship between temperature and output voltage for a particular transducer looks like Figure 4.19. What we need is a relationship of the form

$$\text{temperature} = F(\text{voltage})$$

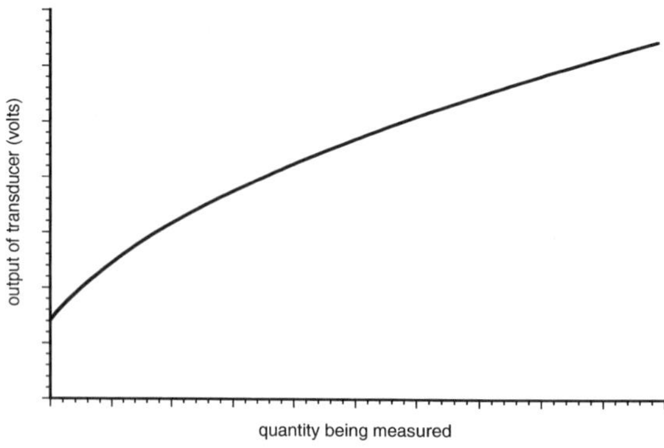

Figure 4.18 A nonlinear transducer: the output does not vary linearly with the quantity being measured.

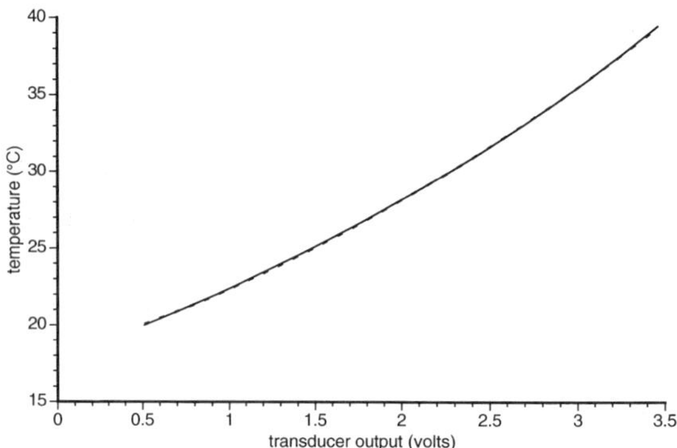

Figure 4.19 This temperature transducer has a nonlinear response. A second order polynomial (dotted line) fits the data (solid line) very well. A voltage from the transducer can be converted to a corresponding temperature using the equation of the polynomial: temperature $= 0.757 \times \text{volts}^2 + 3.54 \times \text{volts} + 18.1$. Note that the output of the transducer (volts) has been plotted on the x-axis so that a standard curve fit of y on x can be used.

Nonlinear curve-fitting programs are available in many standard computer packages such as SAS, SigmaPlot, and DeltaGraph. Jandel also sell a program specifically for this purpose ('TableCurve'). Using such a program a curve that fits the data points can be found. The type of curve (power series, logarithmic, etc.) is irrelevant; for mildly nonlinear transducers a power series is a good starting point. Figure 4.19 shows that a curve of the form

$$\text{temperature} = 0.757 \times \text{volts}^2 + 3.54 \times \text{volts} + 18.1$$

fits the data well. This function can now be used to convert any voltage from the transducer into a true temperature.

The data to plot graphs such as Figure 4.19 usually have to be obtained by experiment. This means that another primary temperature measuring device is needed which may seem to defeat the whole purpose of the exercise. But the primary temperature-measuring device (which may be large and expensive) could be borrowed or rented and used to obtain a calibration curve for a cheap thermistor. The thermistors could then be used to measure temperature accurately in places that the primary device could not be used, in an implanted capsule, for example. Or an array of dozens of thermistors might be used to map out the temperature gradients present across the body surface.

There are a couple of things to remember when using this curve-fitting approach:

1. The calibration curve must span the full range of temperatures to be measured. Do not extrapolate the calibration curve outside the measured range. This is asking for trouble.
2. The conventional way of plotting a graph is with the independent variable on the x-axis. When calibrating transducers, the physical effect is the independent variable and the output voltage from the transducer is the dependent variable. If the data are plotted with voltage on the y-axis and physical effect on the x-axis a standard curve fit gives the relationship in the form

   $$\text{voltage} = F(\text{temperature})$$

   which is not what is required. We need the relationship in the form

   $$\text{temperature} = F(\text{voltage})$$

   so that the voltage coming from the transducer can be converted into the quantity that we want – temperature. Some programs have the facility to fit both the y-axis to the x-axis and vice versa. If not, plot the physical effect on the y-axis and voltage on the x-axis and then perform a standard fit of y on x.

Constructing a complete calibration curve can be time-consuming, and if more than one transducer is being used you either have to calibrate each one or ensure that they are all similar enough that a single calibration curve can be used. In some cases the relationship between the transducer output and the quantity being measured is known. For example, the resistance of a thermistor is a nonlinear function of temperature but is modelled by equations such as

$$R_T = R_{T_0} \exp\left[\beta\left(\frac{1}{T} - \frac{1}{T_0}\right)\right]$$

where R_T is the resistance at temperature T and R_{T_0} is the resistance at a standard temperature T_0 (usually 25°C). Manufacturers can supply the constants β and R_{T_0} for a particular range of thermistors.

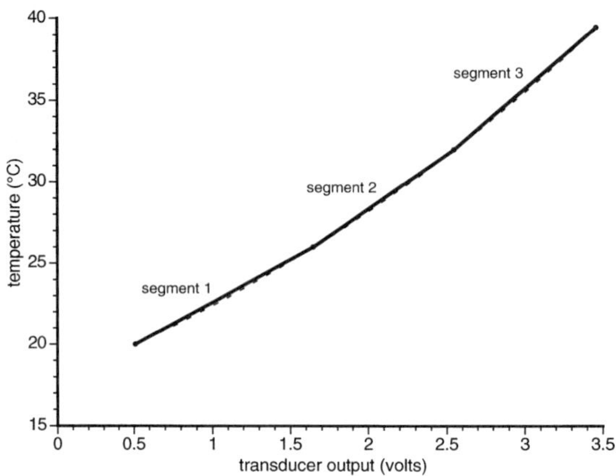

Figure 4.20 Piecewise linearization of the same temperature transducer. The curve is approximated by three straight lines, the equations of which are known.

Another example is a Pitot tube. This is frequently used to measure air velocity. The Pitot tube develops a pressure gradient that is proportional to the square of the airflow velocity, and the pressure gradient is then measured with a pressure transducer. The overall relationship between air velocity and the output voltage of the pressure transducer is thus of the form

$$\text{velocity} = k\sqrt{\text{output voltage}}$$

The constant k can be determined with a single calibration point.

Piecewise linearization is an older linearization technique that is still used occasionally. A calibration curve is constructed as normal, but instead of fitting a curve to the data it is approximated by a series of straight lines (Figure 4.20). The scale and offset of each line is calculated and the data are linearized by determining on which segment of the curve each point lies, then using the appropriate scale and offset to perform a normal linear scaling. It is quite easy to design an electrical circuit that will perform piecewise linearization and thus the nonlinear signal from a transducer can be linearized by such a circuit, which was important before personal computers were readily available. Programming a computer to perform piecewise linearization is straightforward but more complex than the nonlinear curve-fitting method, and the latter avoids the errors introduced by approximating a smooth curve by a set of straight lines.

chapter five

Data Manipulation

Data acquisition systems make it very easy to collect huge amounts of data from many sources at once. The real challenge is in data reduction: obtaining useful information from the mass of numbers. As an example, imagine digitizing a blood pressure trace from an arterial catheter at 50 Hz. The tracing would look something like Figure 5.1 and at a heart rate of 88 beats/min there are thirty-four data points per beat. What we really want is the mean, systolic and diastolic pressure for each beat, plus the instantaneous heart rate: four values per beat.

There are several ways to tackle such problems. At the 'grass roots' level you can write your own programs from scratch in a language such as C, Pascal or BASIC. The next step up is to use libraries of routines that are called by your program. There are libraries in the public domain but these are often older algorithms written in FORTRAN. LabWindows, from National Instruments, takes care of the user interface, graphics, file handling, and so forth whilst still giving you the flexibility of a low-level programming language. Higher-level 'languages' like Matlab, Gauss, Igor, LabView, and HiQ perform all the housekeeping behind the scenes and allow you to concentrate on the data analysis whilst still retaining some programming structure. Higher still are complete acquisition and analysis packages (Acqknowledge, SuperScope) that are designed for people with very little programming skill. They each have their own environment for data analysis and provide a fairly comprehensive library of functions. If this library includes what you want they are wonderful: if not, you are often rather stuck, as adding your own routines to these languages is usually very difficult. As you ascend the scale from programming languages to complete packages you lose flexibility but gain time. Using ready-made routines saves a great deal of time writing and debugging procedures. In general, the quickest way to solve a particular analysis problem is to pick the highest-level system that is flexible enough to do the job.

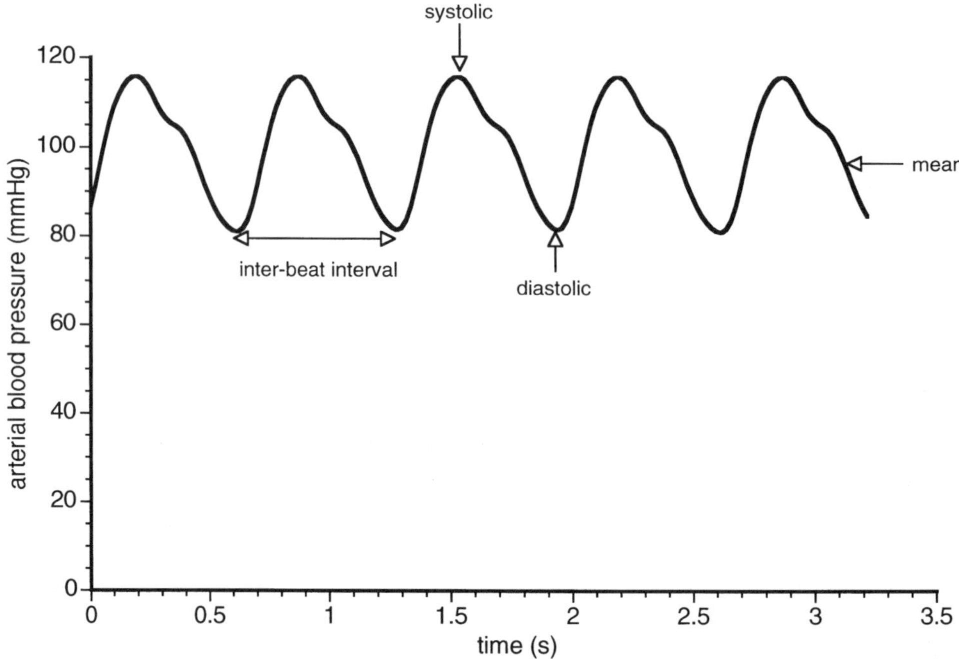

Figure 5.1 Recording of blood pressure from an arterial catheter. The sampling rate was 50 Hz and there are 170 samples in this section of the data. The information that is needed are the systolic, mean and diastolic pressures and the inter-beat interval for each beat, a total of 20 numbers. Obtaining useful information from raw data recordings is one of the most important and challenging aspects of data analysis.

The theory behind manipulating digitized data comes under the umbrella of digital signal processing. The subject is very mathematical but a good introduction to digital signal processing that does not go into too much detail is *Digital signals, Processors and Noise* by Paul A. Lynn. One of the standard books on DSP is *Theory and Application of Digital Signal Processing* by Lawrence Rabiner and Bernard Gold. Approaching the problem from the computing side are the *Numerical Recipes* series (e.g. *Numerical recipes in C: the art of scientific computing* by Press, Teukolsky, Vettering, and Flannery), which have algorithms for solving most scientific computing problems. In addition, the pros and cons of the various techniques for tackling each problem are discussed. The texts and programs are available in C, BASIC, FORTRAN, and Pascal.

In this chapter we explore some of the ways in which data can be manipulated and analysed in order to turn long strings of numbers into useful data points, ready for plotting and statistical analysis.

DATA FORMAT

Data gathered with a data acquisition/analysis system consist of long strings of numbers recorded at regular time intervals. Each channel of the data acquisition unit generates one string of numbers. The strings of numbers have various names: mathematically, they are one-dimensional arrays or vectors. Igor and SuperScope call them waves. In a programming language it is easiest to define them as one-dimensional arrays, but this can lead to problems in some languages if the array contains more than 64,000 data points. We will refer to the strings of numbers as arrays, and assume that they are one-dimensional unless otherwise stated.

STATISTICS

Some of the simplest procedures are statistical descriptions of the data. The *mean* of the samples is very simple to calculate: add up all the samples in the array and divide by the number of samples. When dealing with real-world signals the mean is useful because it reduces noise. Suppose we want to measure the resting potential inside a cell using a microelectrode. The actual value that we obtain from the output of the electrode amplifier will not be absolutely constant but will fluctuate around a value of about $-70\,\text{mV}$. These fluctuations are caused by noise, both within the cell and added by the amplifier. If we take the average over a period of time the size of the fluctuations is reduced and we obtain a better estimate of the true resting cell potential.

The time period over which the samples are recorded is important. How well we reduce the low-frequency components of the noise is determined by how long we sample for, not how fast. By averaging the samples we are low-pass-filtering the signal, which is what we want to do: we want to keep the low-frequency information, which is the resting cell voltage, and reject the high-frequency noise. The frequency response of an averaging filter is shown in Figure 5.2. As we increase the time over which we average, the frequency response of the filter changes and the high-frequency components are attenuated more and more. The first minimum in the response occurs at a frequency equal to 1/(total sampling time), for example, if we sample for $2\,\text{s}$ the first minimum occurs at $0.5\,\text{Hz}$. The longer we average the better we will get rid of the lower-frequency components of the noise. The position of the first minimum is independent of the sampling rate, which may seem odd at first, because if the sampling rate was very low we would only be averaging a few samples. However, remember that the signal we start with is a constant voltage (the resting potential) mixed with higher-frequency noise. We have to sample fast enough so that this high-frequency noise is not aliased.

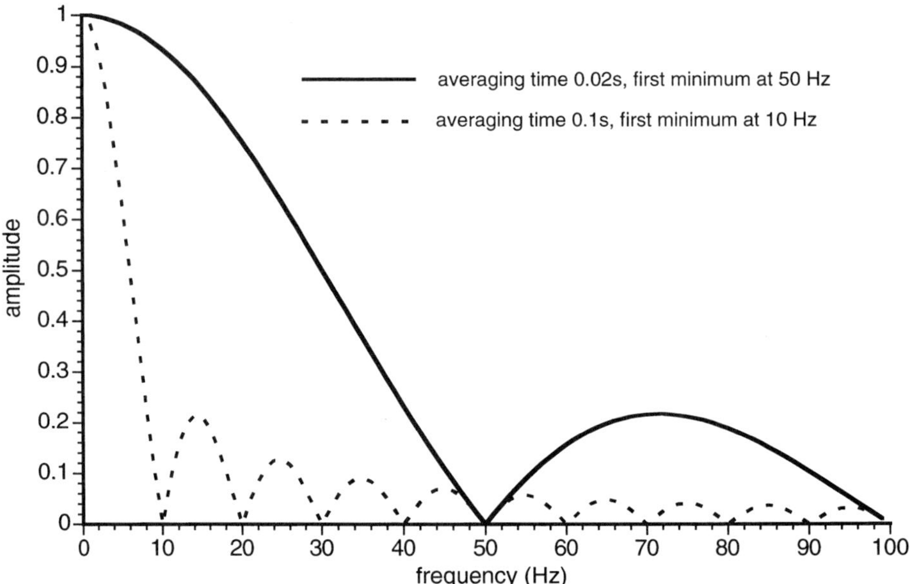

Figure 5.2 Frequency response of an averaging filter. The filter removes the higher-frequency components, and the longer the averaging period the more the higher frequencies are attenuated.

Another very important use of averaging is where the signal that we are interested in is repetitive, or can be made so. As an example, suppose we are interested in recording the electroretinogram (the faint electrical signal produced by the retina in response to a flash of light). A single response is shown in Figure 5.3. Such a low-level signal is usually contaminated by noise from the amplifier and biological noise such as EMGs from the ocular muscles. But we can record many ERG responses and average across the responses to filter out the noise. Averaging across responses means that the first sample in the filtered response is the average of the first sample in each individual response, the second sample in the filtered response is the average of the second sample in each individual response, and so on. The amount by which the noise level is reduced is proportional to the square root of the number of samples averaged, which is typically 100 to 1,000 samples.[1] Averaging across responses works because the noise in one response is uncorrelated with the next and averages to zero. The signal, however, is the same from one response to another and is preserved during averaging. The technique is quite powerful and can extract signals which seem buried in noise. Figure 5.4 shows the original ERG response, the same trace with 100% noise [root mean square (RMS) amplitude of the noise equal to the RMS amplitude of the signal] added, and the result of averaging 100 such traces. The individual responses should all be

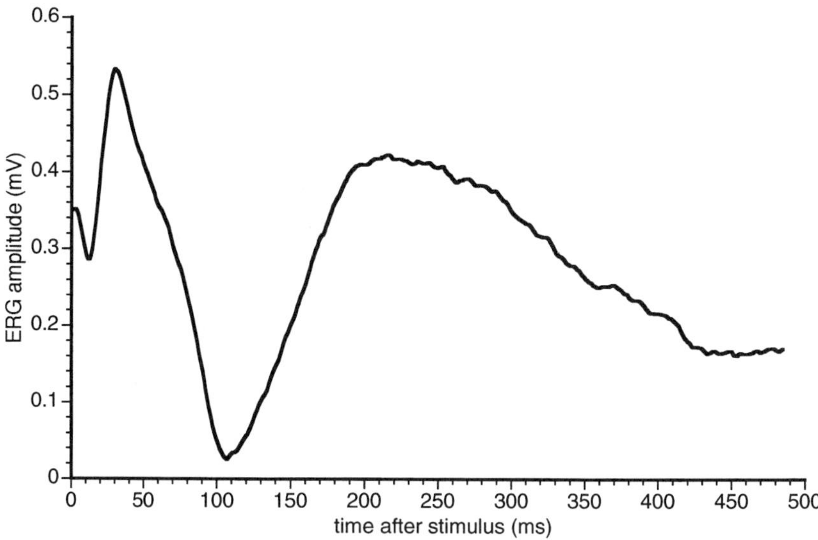

Figure 5.3 The electrical activity of the retina in response to a single flash of light (an electroretinogram).

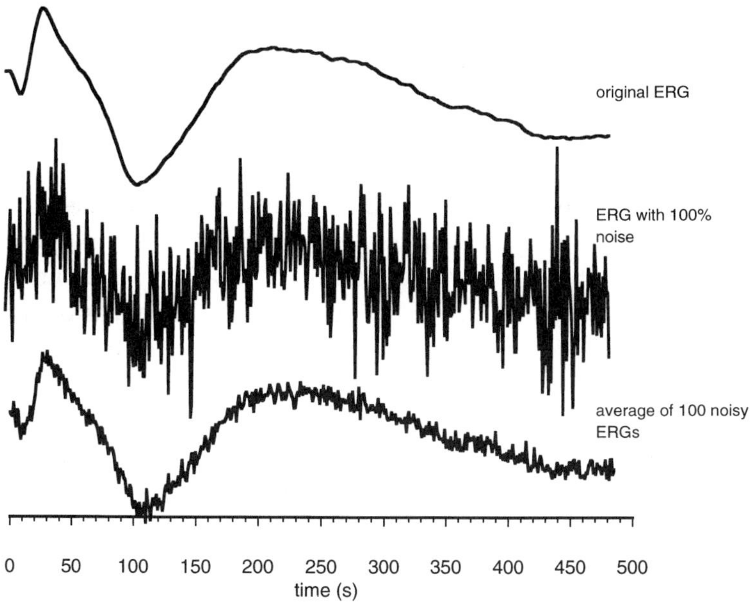

Figure 5.4 Averaging repetitive signals is a powerful technique for reducing noise. The top trace is the ERG recording from Figure 5.3. The middle tracing is the ERG recording with 100% noise added – the ERG signal is almost lost and no detail is visible. The bottom trace is the average of 100 noisy recordings. The shape of the ERG is clearly visible and measurements of the amplitude and latency can be made.

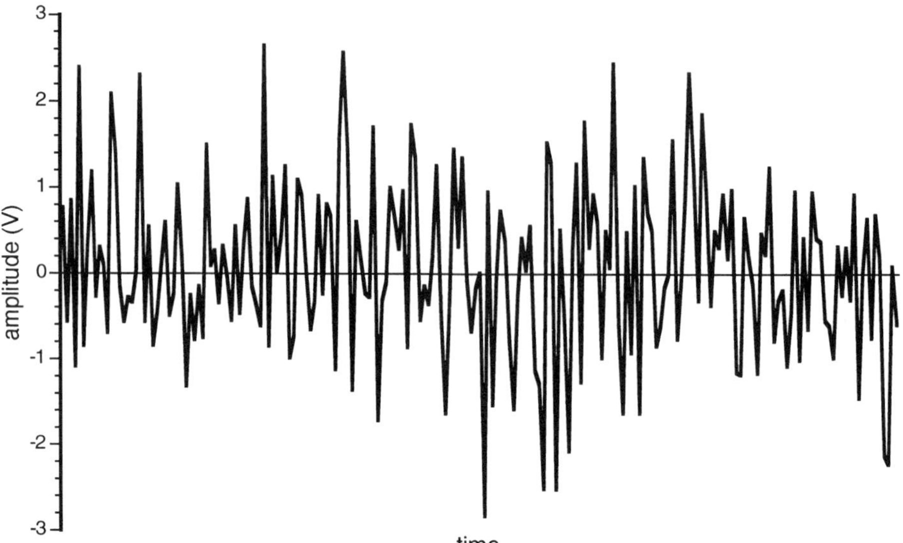

Figure 5.5 A noise-like signal. The RMS voltage is commonly used to describe the amplitude of such noise-like signals.

similar to each other and the number of responses that can be averaged is limited by the time for which the patient can keep still. Nevertheless, signal averaging is a very useful technique in neurophysiology.

A modified version of the mean that is useful for noise-like signals is the RMS value. Figure 5.5 shows a recording of a noise-like signal. Often we need to describe the amplitude of this type of signal but it is difficult to do because there is no pattern to it. This problem occurs, for example, in electroencephalo-gram (EEG) analysis. The EEG signal is noise-like and we need to measure its amplitude somehow so that the effects of drugs on the EEG can be investigated.

The RMS value is calculated by taking each sample, squaring it, taking the mean of all the squared samples and then taking the positive square root of the mean. Just like the simple mean, the signal is low-pass-filtered by the averaging process. Some noise-like signals have frequency components that extend to very low frequencies and with these signals there is no unique RMS value, because the averaging time needed to filter out the lowest frequencies would be too long. What you have to do is trade recording time for some acceptable variation in the RMS value, and this variation can be quantified statistically as the variance or standard deviation. As an example, ten recordings were made of a noise-like signal with long-term mean value of 0 and RMS amplitude of 1 V. The RMS value of each recording was found and the standard deviation of the ten RMS values was

Figure 5.6 Effect of increasing the recording time of a noise-like signal on the variability of its RMS value. The standard deviation of the RMS value of ten recordings is plotted against the recording length (in samples). As the recording length increases the variability of the RMS value falls because of the low-pass filter effect of averaging over the whole recording length.

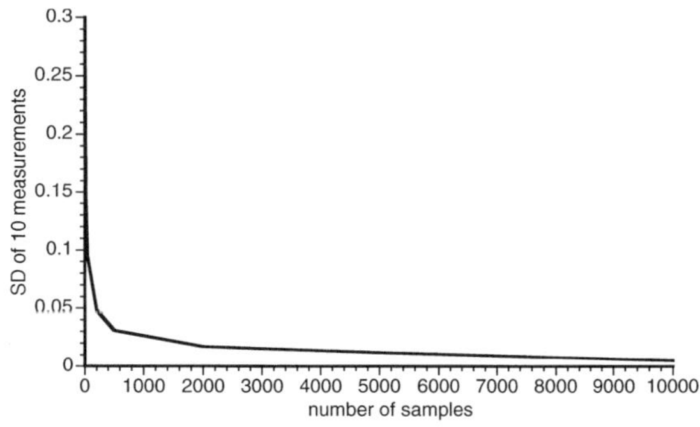

Figure 5.7 Four quite different signals with the same RMS amplitude (354 mV).

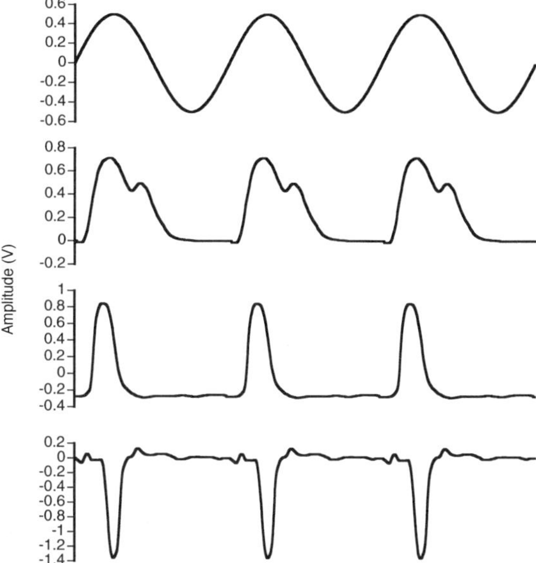

calculated. The process was then repeated with increasing recording lengths. Figure 5.6 shows the standard deviation of the ten RMS values plotted against recording length.

RMS values can be calculated for any type of waveform, not just noise-like signals. If the signal is a regular waveform with no random components then there is a unique value of the RMS amplitude. In fact, the RMS amplitude is the 'universal' way of describing how big a signal is. Figure 5.7 shows a variety of signals with very different shapes. If we want to compare the amplitudes of signals of the same type there is no problem: we can just pick certain points on each waveform and measure how large they are. However, if we want to compare the four different signals in Figure 5.7, we have a problem, because the waveforms have different shapes. The RMS value of a signal is a way of describing the size of the signal regardless of its shape. The RMS value also has another property: it is a measure of the energy in each signal. If we have a signal with an RMS amplitude of 250 mV then we know that whatever its shape, if we connect the signal across a resistor we will get the same amount of energy from it. The amount of energy released in the resistor is $W = V^2/R$, so if the resistor has a value of $100\,\text{k}\Omega$ then the power dissipated in it will be $0.625\,\mu\text{W}$. All the signals in Figure 5.7 in fact have the same RMS amplitude (354 mV) although their peak amplitudes vary considerably. Note that if a signal consists of a constant (DC) component plus a varying (AC) component, both contribute towards the RMS value.

PEAK, TROUGH, AND ZERO-CROSSING DETECTION

Very frequently, we need to know the value of the highest point (peak) or lowest point (trough) in a recording. Figure 5.8 shows a single motor evoked potential (MEP) recorded from a dog. MEPs are characterized by their peak to peak amplitude and latency (the time from the stimulus to the first peak). The simplest way to identify the largest peak and trough is to find the largest and smallest samples, respectively, However, this is not the best approach because it is susceptible to noise and is not extendible to repetitive waveforms.

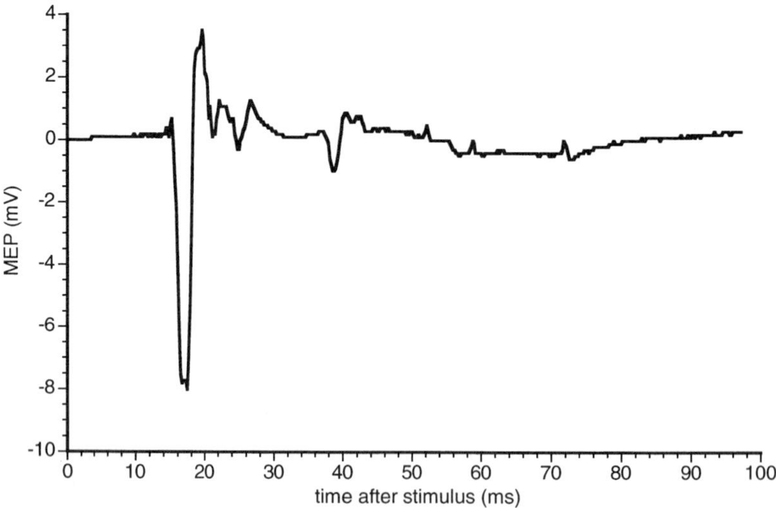

Figure 5.8 A single motor evoked potential recorded from the anterior tibialis muscle in an anaesthetized dog. The motor nerve to the muscle was stimulated non-invasively using a magnetic coil.

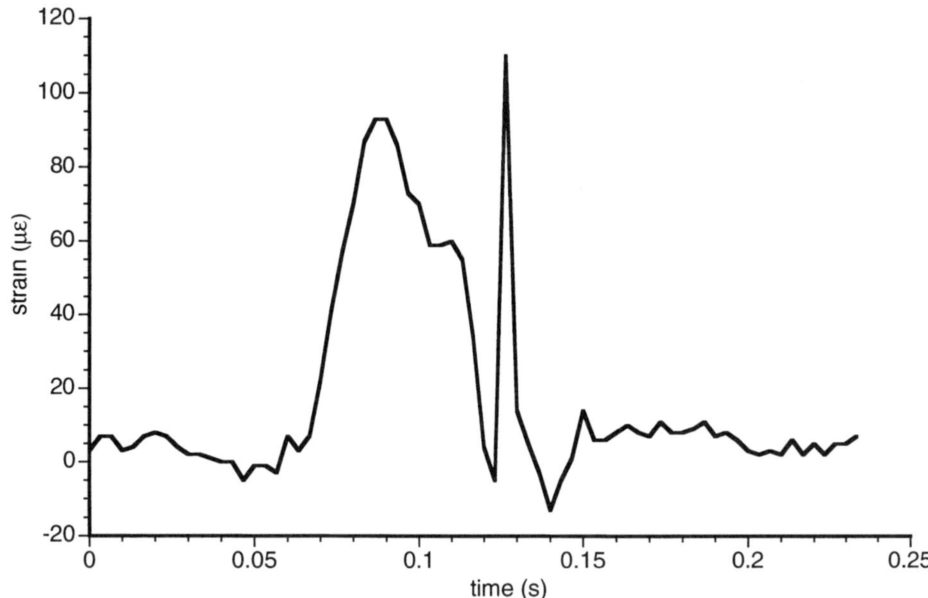

Figure 5.9 A recording of the deformation of a horse's hoof during a single stride. Deformation is measured in microstrain ($\mu\varepsilon$). A large noise spike occurs at 0.13 s, just after the main impact. Signal-processing algorithms must be able to recognize and reject common artefacts such as this.

Noise causes problems in all peak-detection methods. Figure 5.9 shows a pulse whose height and position in time we would like to measure. Like all biological signals, it is contaminated with noise and there is a large spike after the main peak. If we just pick the highest point we will miss the true peak and both the height and position will be inaccurate.

The first step in dealing with the problem is to low-pass-filter the data to remove the components of the noise that are higher in frequency than the signal. However, the second problem still remains: if we have a string of pulses (Figure 5.10) we usually want to record the height and position of each one, so we need an approach that detects *local* minima and maxima. Mathematically, maxima and minima are points on a curve at which the slope of the curve is zero. We can easily calculate the slope of a signal by differentiation (see below). But this emphasizes any high frequency noise present, which is what we are trying to avoid.

In the end, the procedures used for detecting local maxima and minima are often quite ad hoc and are tailored to each particular signal type. A rule-based system is commonly used: when various criteria are fulfilled then a peak/trough is recorded. Some of the tricks used are:

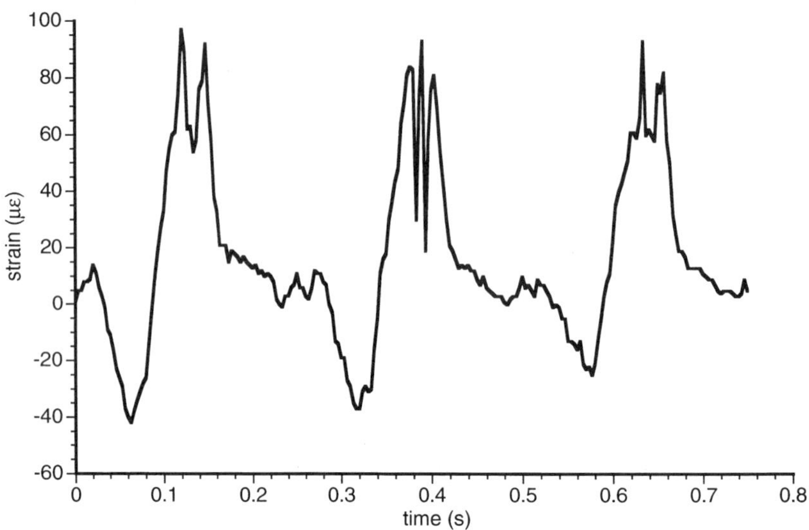

Figure 5.10 A longer recording of hoof wall deformation that spans three strides. Noise spikes at peak deformation are clear. A signal-processing algorithm is needed that will filter out the spikes and then find the height and position of each of the three peaks.

1. Make decisions based on a running mean rather than single values. The running mean (typically 3–5 samples) acts as a low pass filter and reduces the effects of noise.

2. Calculate the local slope of the curve by fitting a curve to it over a few points, then calculating the slope of the curve. Quite a modest curve fit will do. Again, this tends to reject noise rather than accentuate it as true differentiation would do. The peaks and troughs can then be found by looking for the point at which the slope changes from positive to negative or vice versa.

3. Define a *hysteresis* band, which is a sort of dead zone. In order for a peak to be detected, the signal must rise out of the top of the band and, once a peak has been detected, another will not be recorded until the signal has fallen out of the bottom of the band (Figure 5.11). This means that large peaks are detected but small wiggles are rejected. The use of a hysteresis band is a powerful way of detecting pulses in the presence of noise. The process is the digital version of the *Schmidt trigger*, an electronic circuit that uses a hysteresis band to detect pulses reliably that are mixed in with noise. The hysteresis band is typically 3–5 times wider than the amplitude of the noise present. The upper and lower limits of the hysteresis band can either

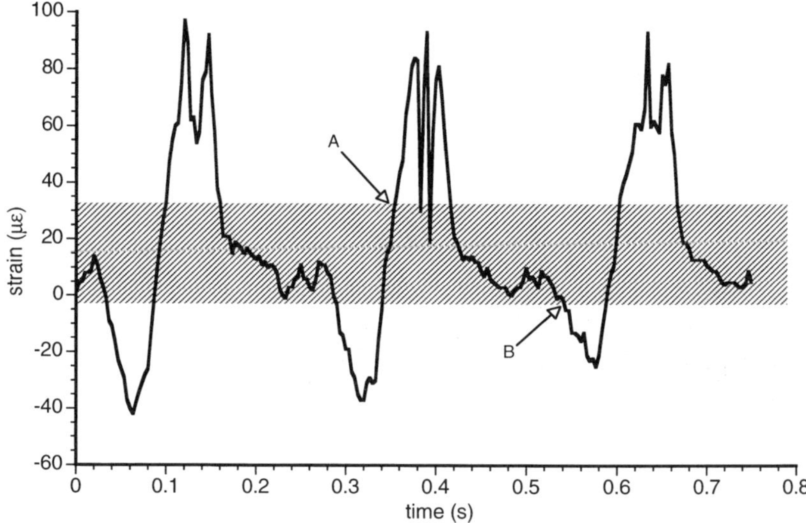

Figure 5.11 A hysteresis band is a very good way of rejecting noise during peak detection. The hatched area is the hysteresis band. Once the signal rises above the high side of the band (point A), a peak will not be recorded until the signal drops below the low side of the band (point B). Thus the algorithm will correctly identify only one peak between points A and B in spite of the large noise spikes.

be fixed values or calculated from the signal itself. For example, the upper threshold could be set to be 50% of the average height of the last 20 peaks. This is an example of an *adaptive* process, that is, one that changes according to the nature of the signal presented to it. Adaptive processes have to be designed carefully to make sure that they are stable and start up properly.

4. Characterize each peak/trough and reject ones which fall outside some acceptable range. For example, a big noise spike might be large enough to exceed the hysteresis band and thus will be detected as a legitimate peak (Figure 5.12). However, the width of the pulse may be much less than a real pulse, and so such noise spikes can be rejected by comparing their widths to a predetermined range. Similarly, the time between pulses can be examined and pulses which are too close together rejected. An example would be heart rate detection from the ECG signal, a very common problem in medical signal processing. The normal heart rate of an adult human will be within the range of 30–210 beats/min, so the time between beats must be between 2.00 s and 0.29 s. Interbeat times outside this range should be rejected.

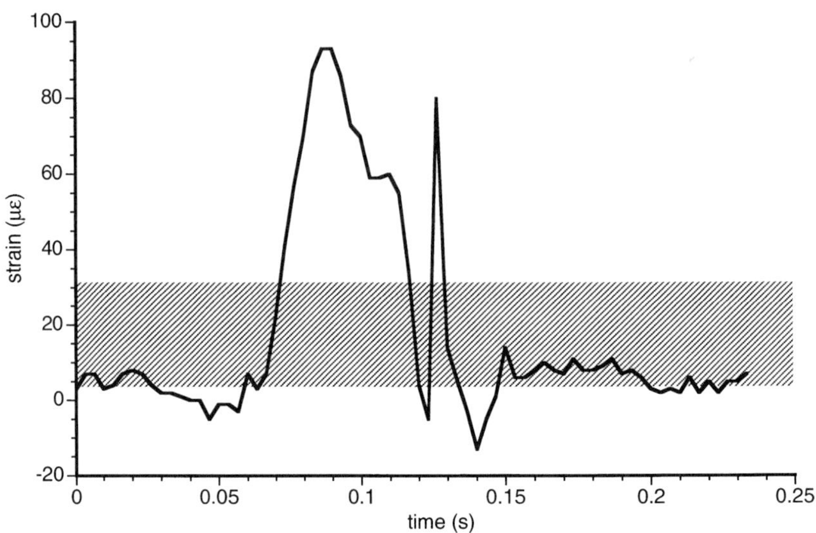

Figure 5.12 A very large spike can exceed the width of the hysteresis band and be incorrectly recorded as a peak. Further information can be used to reject such spikes. The spike is very narrow (about 0.01 s) and is very close to the previous peak, distinguishing it from a genuine peak. Low-pass-filtering the data before peak detection would also help by reducing the amplitude of the noise spike.

Specific rules such as the interbeat distance cause problems when an unusual signal is processed. For example, small mammals and birds have heart rates that are much higher than the normal human range and can exceed 400 beats/min. When ECG monitors designed for use on humans are used on animals, the heart rate displayed on the front of the monitor is often completely wrong. The ability of a program to reject noise and artefacts and yet deal with a range of signals is known as its *robustness*.

A task related to detecting peaks and troughs is detecting *zero crossing*, the point at which a signal changes from positive to negative or vice versa. Zero crossing detection is the first step in analysing respiratory flow signals because the points of zero flow mark the start of inspiration and expiration. As with peak detection, simply examining the sign of a signal is of no use because noise will make the signal cross zero several times at each transition (Figure 5.13). The tricks used to write a robust zero crossing detector are the same as those used in peak/ trough detection. In particular, a hysteresis band which sits around the zero mark (Figure 5.14) is used. Once the signal breaks out of the top of the hysteresis band it is regarded as being positive until it breaks through the bottom of the hysteresis band, so that small dips in the waveform are ignored.

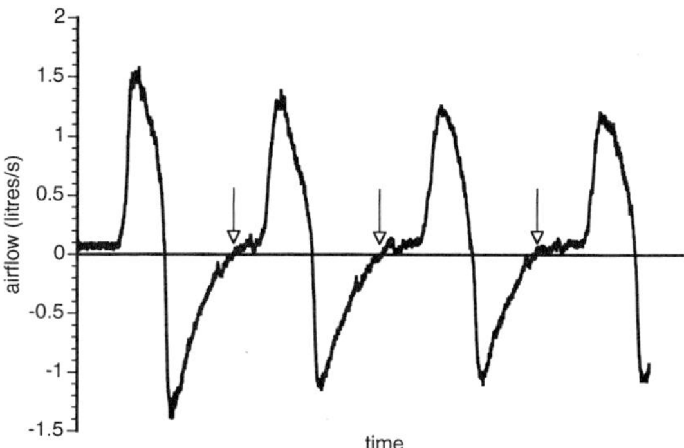

Figure 5.13 Zero crossing detection. Arrows mark the places where the flow signal from a pneumotachograph crosses zero. The small amount of noise present with the signal means that it actually crosses and re-crosses the zero line several times during each transition so that simply looking for changes in the sign of the flow signal will result in many false positives.

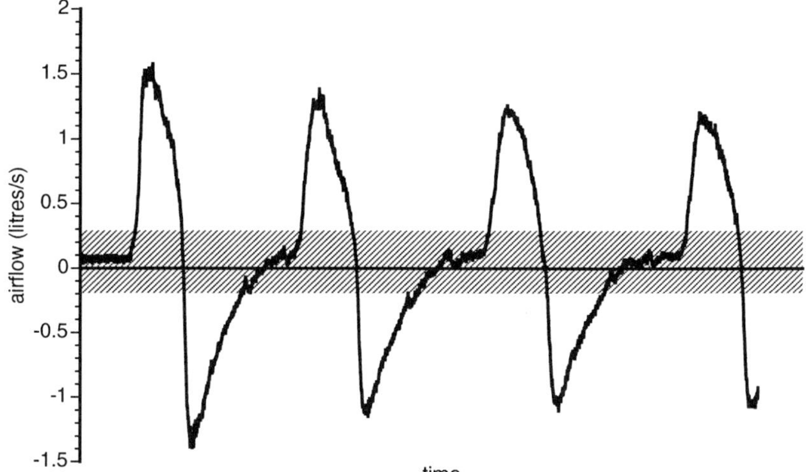

Figure 5.14 Adding a hysteresis band (hatched area) to the flow signal eliminates the noise problem but the precise moment of zero crossing is not known. It is relatively easy to obtain a good estimate of the true point of zero crossing.

The hysteresis band introduces some errors in timing because the precise moment of zero crossing is not known. There are a couple of ways to avoid the problem:

1. Ignore it. If there is not much noise and the hysteresis band is narrow, the error is small.
2. Linearly interpolate between the two data points where the signal enters and exits the hysteresis band (Figure 5.15). In effect, a straight line is drawn between the two points and the place where the line crosses zero (which can be calculated precisely) is taken to be the zero crossing point.

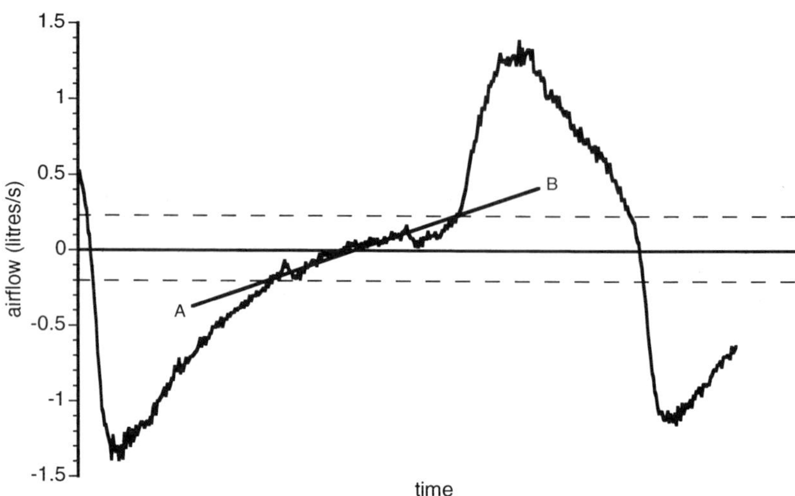

Figure 5.15 One way of estimating the true zero crossing position. The points at which the signal enters and leaves the hysteresis band (upper and lower limits of the band are shown by dotted lines) are recorded and a straight line (A–B) drawn between them. The point at which the line crosses zero can be calculated precisely and is a good estimate of the true zero-crossing point of the flow signal.

The start and end of arrays play havoc with many signal-processing algorithms including peak/trough and zero-crossing detectors. Although one does not expect to get a sensible reading from a pulse that has been chopped in half at the end of an array, it is important to make sure that the program knows this and is not fooled by such artefacts (or even worse, crashes!). Functions such as filtering need several consecutive points on which to work and clearly run into problems at the start and end of a recording. There are no hard and fast rules on what to do. Extending the array as far as necessary with zeros ('*zero padding*') works in many cases. If the array has a mean value that is not zero, it may be better to pad the array with the mean value because this reduces the size of the 'step' between the padded values and the real data. Whatever technique is used, some error is inevitably introduced and it is sensible to collect data in long blocks so that these errors are reduced as much as possible.

Collecting data in single long blocks is often not desirable and in many cases may not be practical. For example, suppose we are interested in the changes in respiratory pattern of an anaesthetized animal during a four-hour experiment. We need to collect respiratory flow data continuously and analyse it to get tidal volume, inspiratory and expiratory times, and so forth for every breath. We could collect the data in one large block and process them at the end of the experiment, but it is far better if the data can be processed and displayed whilst the experiment is in progress. This means that the computer has to collect data and process them at the same time. Most data acquisition systems can do this, but in order for the computer to still be responsive the data usually have to be collected in small blocks less than a couple of seconds long. The computer processes the last block of data while the next one is being collected. The challenge is now to process the blocks of data as if they were all one long record. Conventional techniques that ignore partial breaths at the ends of arrays are no good because each block of data probably contains less than one breath. The solution is to write analysis programs that remember their current state (the values of all the variables that they use) at the end of a block and start off using the same values at the beginning of the next block – in effect, picking up where the program left off. If the algorithm has to operate on several data points at once then the last few data points in the block are also stored inside the program and added to the beginning of the next block.

RESAMPLING AND INTERPOLATION

Even the most robust peak detection program still has one weakness: the exact position of a peak is not known. Simple peak-detection algorithms find the largest sample in a small subset of the data that contains the peak. These algorithms therefore find the sample nearest to the true signal peak and not the peak itself. Figure 5.16 shows that both the position and amplitude of a peak are approximated by the position of the nearest sample: it is very unlikely that a sample will land exactly on the top of a peak. Even though the signal is sampled adequately (i.e. at a rate higher than twice the maximum frequency present) the error associated with missing the true peak may be unacceptable. One answer is to *oversample*, that is sample at a rate much higher than the minimum required by the Nyquist frequency. However, one can quickly run into problems of data storage when sampling at high rates.

An alternative is to *resample* the data at a higher rate than it was recorded at. This is allowed by Nyquist because we are increasing the sampling rate, not decreasing it. We do not gain any extra information but we can predict the true position and amplitude of the peak more accurately.

Resampling involves placing extra samples in between those originally recorded. Adding one point in between each existing point doubles the sampling rate, adding two (evenly spaced ones) triples the sampling rate, and so on. The amplitude of each point can be determined in a number of ways (see later) but for now we will assume that they are determined by linear interpolation, that is, a

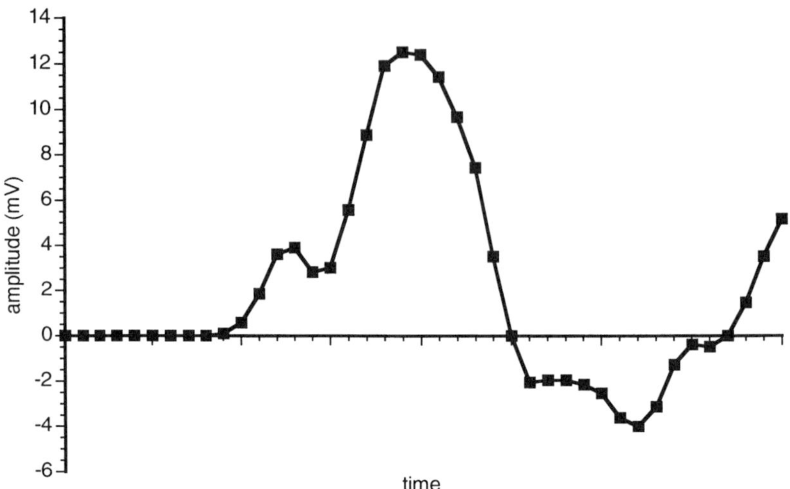

Figure 5.16 A continuous signal (solid line) and its digitized version (squares). Peak-detection algorithms that work on the digitized signal will find the sample that is nearest to the peak of the continuous signal, not the true peak itself. It is very unlikely that a sample will land exactly on the true peak.

straight line is drawn between the original samples and the new samples are evenly spaced on this line. The next step in the process is to low-pass-filter the resampled signal. Adding the extra samples introduces high-frequency artefacts not present in the original signal. Remember the 'spiky' rule: the sharp corners in the resampled signal contain high frequencies that must be removed. The corner frequency of the filter can be at least as low as the original Nyquist frequency without losing any data. Figure 5.17 shows a pulse being resampled (by linear interpolation in this case) at five times the original rate and smoothed to define the peak better. For the smoothing operation a low-pass filter with a corner frequency equal to half the original (not the resampled) Nyquist frequency was used.

The actual method used to determine the amplitude of the added samples is not that important because the low-pass filter will reset the new samples to their correct values. However, the further the samples are placed from their final positions the greater the demands placed on the filter because the high-frequency artefacts have larger amplitudes. If interpolation is computationally too intensive then repeating the previous sample is a sensible alternative.

Sometimes we need to resample a signal at a frequency lower than the one at which it was recorded. Now we have to remember that we are lowering the Nyquist frequency, so the signal must be low-pass-filtered to remove any components greater that half the new sampling rate. It is easiest if the new sampling rate

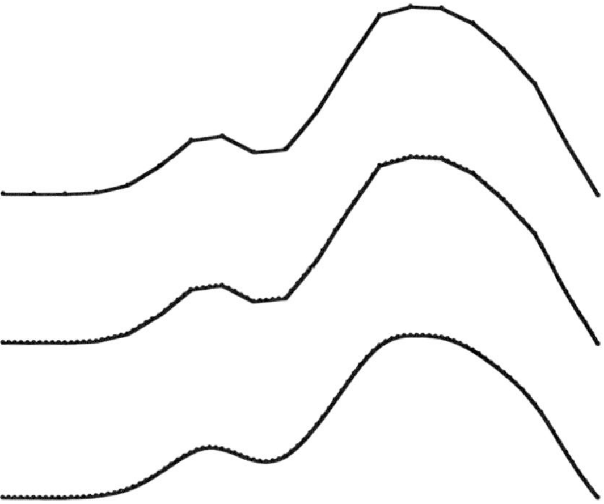

Figure 5.17 Resampling. The top trace is a section of a sampled signal. The samples themselves are indicated by small circles, the lines are for clarity only. The middle trace is the same signal resampled at five times the original rate using linear interpolation. The sharp transitions at the original sampling points introduce spurious high-frequency components. The lower trace is the resampled signal after low-pass-filtering. Note that although the lower trace appears smoother it contains no more information than the top trace.

is an integer division (one-half, one-third, etc.) of the original rate because then the resampling can be performed simply by choosing every second or third sample from the original sequence. Noninteger sampling rate reductions are performed by adding the extra samples (using interpolation or repetition of the last sample), low-pass-filtering and then rejecting the unwanted samples.

CURVE FITTING

Once a set of data points have been cleaned up and plotted, we often wish to draw a single smooth line through the points. This may just be for aesthetic purposes, but often the aim is to gain further information from the data. An example would be in a drug elimination study. The elimination kinetics of the drug can be determined from the plot of drug plasma concentration versus time. There are programs dedicated to particular curve-fitting tasks such as drug kinetics and quantifying peaks on chromatograph results. However, for many studies we need a general-purpose tool to help analyse the data. The aim is fairly simple: we need a means of finding the function that fits our data points optimally in some way. The mathematics behind the subject is quite involved (see *Numerical Recipes in C*, Chapter 15) and we will not go into it here, but will just show how the techniques can be used. In spite of the mathematical techniques used, curve fitting is not an exact science and there is some 'art' to it.

We start with a graph of the data. As an example, Figure 5.18 shows the oxygen consumption of a horse exercising on a treadmill plotted against treadmill speed. Next, we need to choose a mathematical function that we think might fit the data

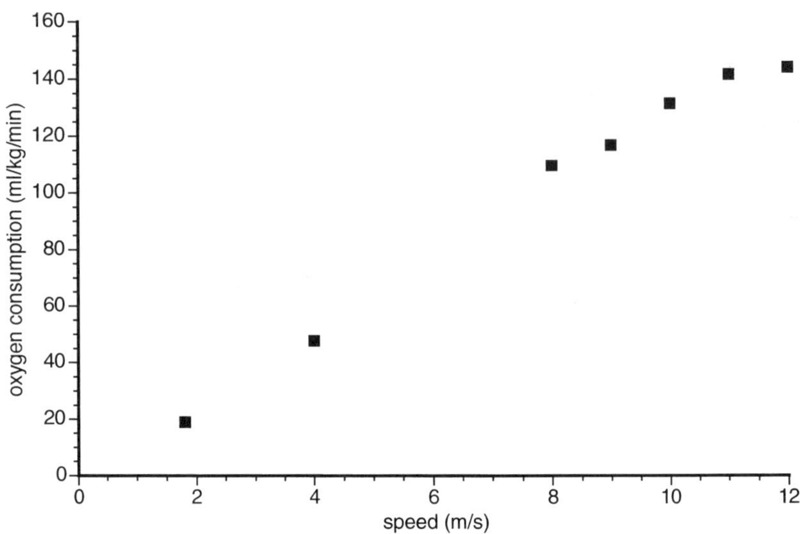

Figure 5.18 A plot of the oxygen consumption of a horse running on a treadmill versus treadmill speed.

points, that is, we need to describe the relationship between the y and x values of the data points in some way. At this stage we usually have one of two objectives:

1. The relationship itself is not important, we just need a nice smooth line through the data points.
2. We know what the relationship is and need to find the parameters that best fit the observed data.

With situation (1), we need to choose a suitable function that looks like the graph. Many programs have the ability to fit basic functions to data sets: in addition to data analysis programs (Matlab, LabView, etc.) most statistical and graphing programs will do curve fits on simple functions. If the graph forms a straight line or a smooth curve then there are several functions that can be fitted to the data:

a. a straight line,
b. exponential increase/decrease,
c. logarithmic increase/decrease, or
d. polynomial. The 'degree' or 'order' of the polynomial determines how many wiggles the fitted curve can have. A polynomial of degree n can have up to $n-1$ peaks or troughs in it (a polynomial of degree 1 is the same as a straight line).

Figure 5.19 shows the same plot of oxygen consumption against treadmill speed, with various functions fitted to it. The third order polynomial fits the data points quite well. Figure 5.20 shows the same data with a cubic spline curve fitted to it. Cubic splines produce a smooth curve that passes through every data point. Note the 'kink' in the curve at 9 m/s. There is clearly some noise associated with the data, and forcing the curve to go through every point is not ideal in this case. A better result is obtained with the polynomial curve that gives a line of best fit rather than going through every point.

A more important use of curve fitting is situation (2), where we have some idea of the relationship behind the data and need to find the set of coefficients that best fit the data set. For example, first-order drug elimination is described by the equation

$$C_t = C_0 e^{-kt}$$

where C_t is the plasma concentration at time t, C_0 is the initial plasma concentration, and k is the elimination constant. Figure 5.21 shows the plasma lactate concentration at various times during recovery from a maximal exercise test in a horse. An exponential curve of the form

$$y = ae^{-bt}$$

fits the data very well ($r^2 = 0.999$) and we can conclude that first order kinetics are a good model for lactate removal under these conditions. The equation of the fitted curve is

$$y = 8.43e^{-0.0558t}$$

The value of b (0.0558) is useful because it allows us to calculate the plasma half-life ($t_{1/2}$) of lactate from the equation

$$t_{1/2} = \frac{0.693}{b}$$

giving a half-life of 12.4 min in this case.

Historically a great deal of emphasis was placed on trying to force the data into some sort of straight line, because a straight line can be fitted easily to a data set by linear regression. Thus data which fitted an exponential process were always transformed by taking logs to fit a linear relationship. It is not really worth the effort of transforming data now because any curve can be fitted directly to the data set.

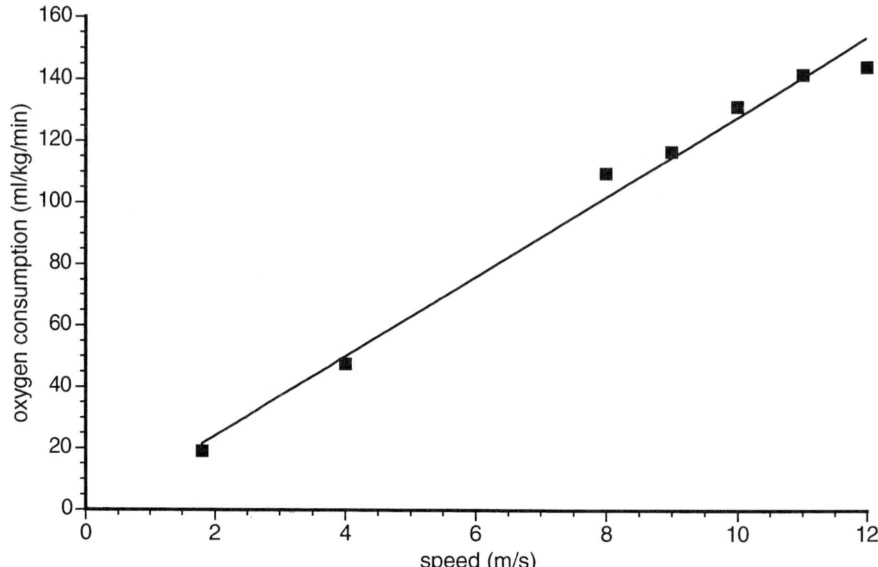

f(x) = 1.295305E+1*x + -1.755759E+0
R^2 = 9.872661E-1

(a)

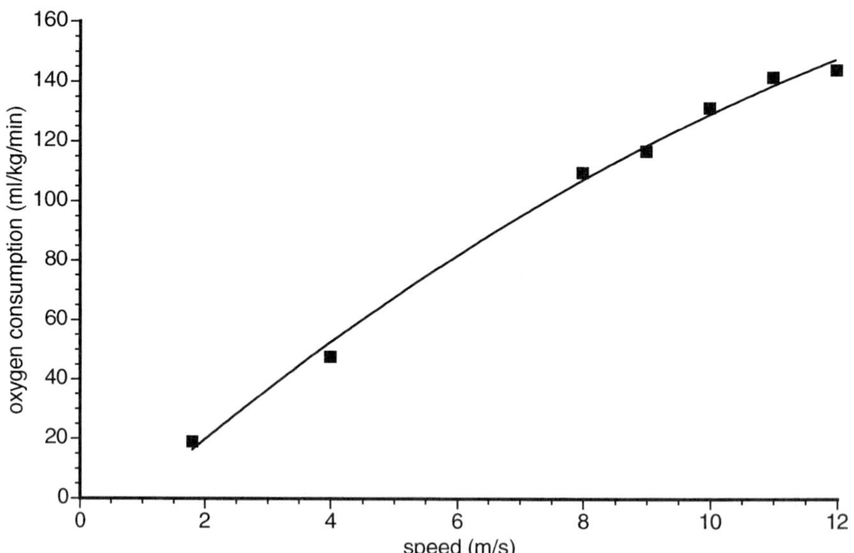

f(x) = -4.435562E-1*x^2 + 1.899974E+1*x + -
1.642113E+1
R2^2 = 9.177415E-1,R1^2 = 9.872661E-
1,R0^2 = 9.954338E-1

(b)

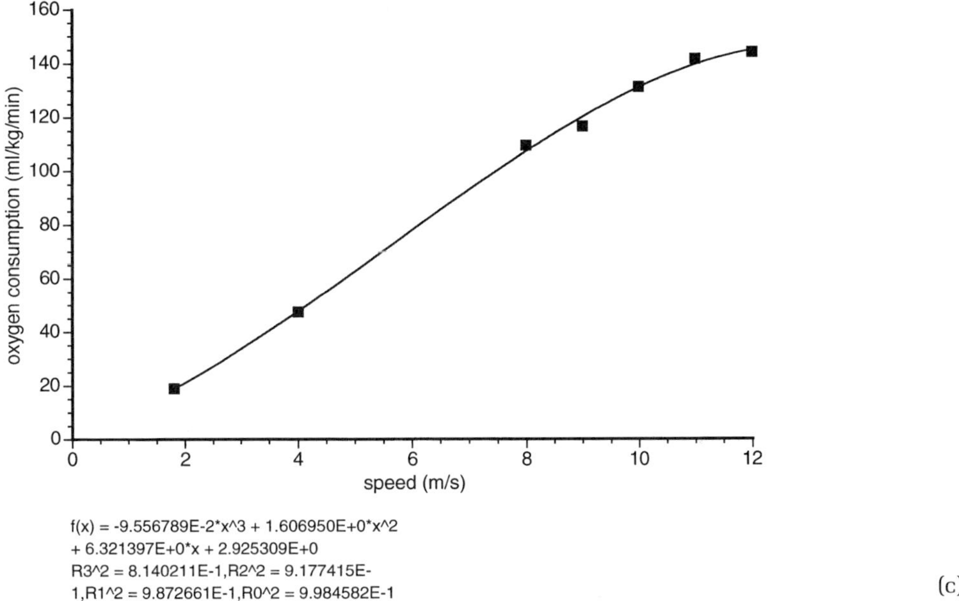

(c)

Figure 5.19 The same data as Figure 5.18 with three different functions fitted to it: (a) a straight line (also called a linear curve fit or linear regression). The equation of the line is: $y = 12.95x - 1.76$; (b) a second-order polynomial. The equation of the line is $y = -0.44x^2 + 19.0x - 1.64$; (c) a third-order polynomial. The equation of the line is $y = -0.096x^3 + 1.61x^2 + 6.32x + 2.93$.

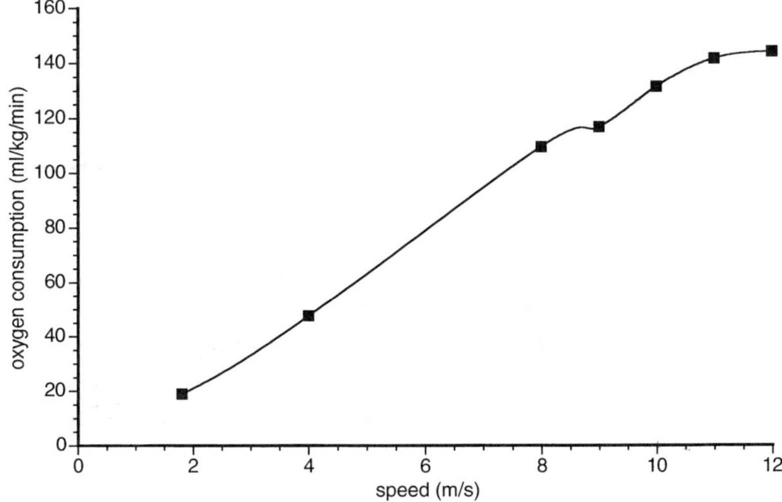

Figure 5.20 The data from Figure 5.18 with a cubic spline fitted to the points. A cubic spline has to go through every data point and the resultant curve is not very useful in this case.

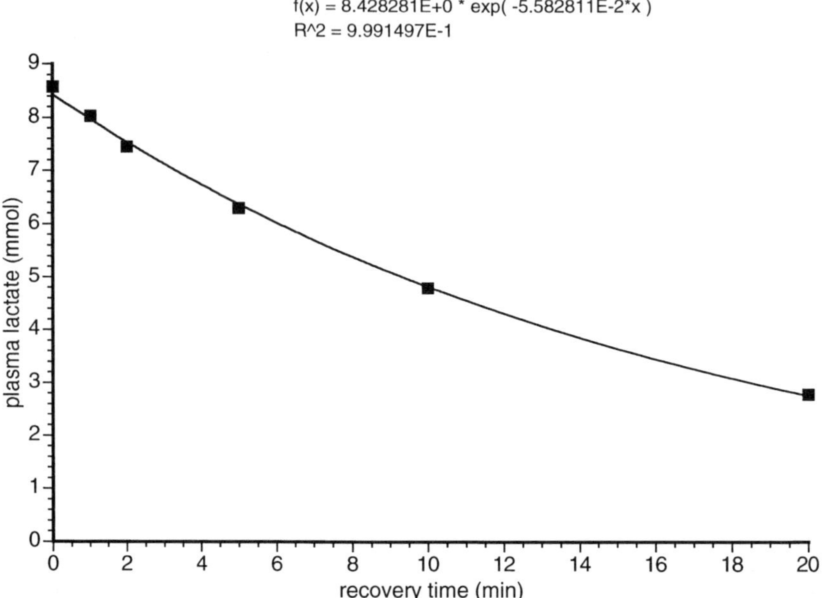

Figure 5.21 Plasma lactate concentrations in a horse recovering from strenuous exercise. An exponential curve fits the data very well ($r^2 = 0.999$). The equation of the curve is $y = 8.43e^{-0.0558}t$, where y is plasma lactate concentration and t is time.

An important question to ask when using curve-fitting routines is 'how well does this curve fit the data?'. For polynomial fits the regression coefficient (r or r^2) is calculated. But for general curve fitting different 'goodness-of-fit' indices are used. One of the most common is the sum of squares. At each data point the difference between the y value of the real data point and the y value of the fitted curve is calculated and squared. The squares are summed to give an overall indication of how well the curve fits the data set: the smaller the sum of squares, the better the fit. Since the aim of the curve-fitting process is to minimize the sum of squares the process is usually called a least-squares fit. One of the entrenched dogmas of curve fitting is that if you do not know the relationship underlying the data you should start with the simplest curve possible (a straight line) and only use more complex functions if they fit the data better. Searching for relationships in data sets without some *a priori* knowledge of what you are after is often called 'dredging'. The prize for the ultimate data dredging program goes to Jandel's TableCurve. This program takes a data set and throws hundreds of functions at it in turn, ranking each one according to how well it fits the data.

Several computer packages will do curve fitting of complex functions, including DeltaGraph, TableCurve, SigmaPlot, Origin, Igor, and Prism. All of them have a library of built-in functions, and most allow you to type in any equation that you wish to fit to the data. Only a few functions can be fitted directly to data using an analytical method.[2] Complex functions are fitted with an iterative method (the Levenberg–Marquardt algorithm is commonly used). The program is given the function to fit, which will have a number of variables in it (a and b in the lactate example above). The program starts with some values for a and b, draws the curve of the function using these values and examines how well the curve fits the data using the residuals. From the results the values of a and b are updated (this is the difficult bit!) and the curve recalculated. As long as the values of a and b are changed in the right direction the function will become a better and better fit to the observed data. Eventually, no amount of tweaking a and b will improve the curve fit and the program stops, giving you its best estimates for a and b and the final goodness of fit. Figure 5.22 shows the total respiratory reactance of a horse over the frequency range 1.5–5 Hz. From theory the reactance can be described by the function

$$y = 2\pi f I - \frac{1}{2\pi f C}$$

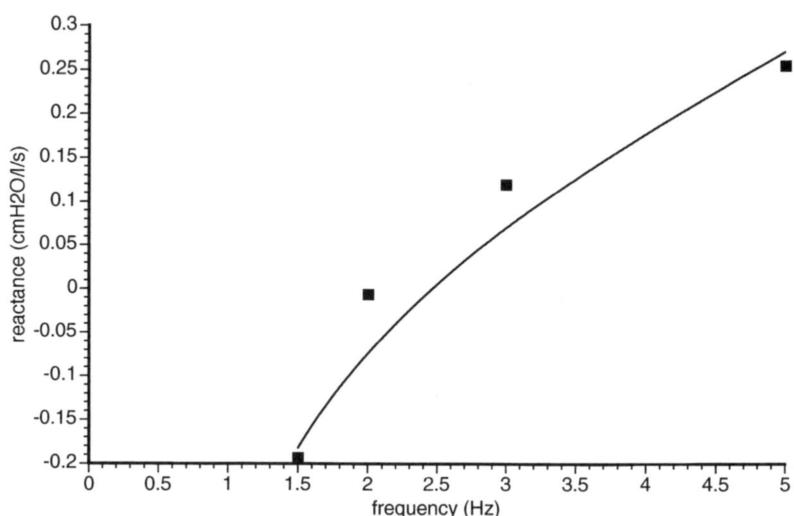

Figure 5.22 A graph of total respiratory reactance of a horse plotted against measurement frequency. From theory the shape of the curve is $y = 2\pi f I - 1/(2\pi f C)$, and this equation was fitted to the data points using the Levenberg-Marquardt algorithm. The values of I and C can be found from the resultant curve fit.

where y is the reactance and f the frequency. The variables (also called *para-meters*) I and C are the inertance and compliance, respectively. This function has to be fitted iteratively. The graph shows the best fit of this function to the data ($r^2 = 0.936$, not a bad fit) and the values of I and C are $0.011\,\mathrm{cmH_2O}\,\mathrm{l}^{-1}\,\mathrm{s}^2$ and $0.37\,\mathrm{l\,cmH_2O^{-1}}$, respectively.

Several things can go wrong with iterative curve fitting. Firstly, the initial values for a and b need to be sensible. If the initial values are way out (e.g. by an order of magnitude) then the program may go off on a wild goose chase and never find the 'right' values for a and b. Most programs allow you to enter initial values for the variables, and it is worth taking the time to make sure these are at least in the right ballpark. Secondly, the program may think it has found the best fit but is actually stuck in a *local minimum* – a set of values that fit the data quite well and appear to be the best solution but are not. One way to avoid this is to run the program several times on the same data set with different initial values. If the final values are all very close to each other then it is likely (but not certain) that the best fit has been found. Finally, it is also important to remember that iterative methods do not guarantee that the best fit has been found, unlike analytical methods. There may not be a unique solution and it is quite possible, especially with noisy data, that there are several sets of values for the coefficients all of which fit the data equally well. For those interested in the details of curve fitting and data modelling, Chapter 15 in *Numerical Recipes in C* is a good start.

FILTERS

Some of the different types of filters, and how to define their properties such as corner frequency and roll-off rate, were mentioned in Chapter 3. We can make filters out of electronic components (hardware filters) that act on electrical signals. But we can do the same after the data have been digitized, using software filters or *digital* filters. A digital filter is a program that takes an array of data and performs exactly the same function as a hardware filter – it removes some frequency components and not others. Whether we use a hardware filter and then digitize the signal, or digitize the signal and then use a digital filter, the end result is the same. There is, however, one place where a hardware filter is essential and that is the anti-aliasing filter (Chapter 3). All frequency components above half the sampling rate must be filtered out of a signal before it is digitized. The low-pass filter which does this has to be a hardware filter because the signal is still in analogue form. Once the signal has been sampled, however, you can go wild with digital filters!

Digital filters have some important advantages over analogue ones:

1. They cost nothing to construct except a little time. All the main data acquisition/analysis packages have built-in digital filter functions which are very easy to use.
2. You can construct filters which are impractical to make in analogue form. High-order filters have many components and need careful shielding between stages. Analogue filters that operate at low frequencies (e.g. with a cut-off frequency of less than 0.1 Hz) need bulky and expensive capacitors.
3. Digital filters are stable over time. Analogue filters 'drift' because the component values change over time. The 'components' in digital filters are numbers (*coefficients*) which do not change.
4. Digital filters can be matched precisely because the same coefficients can be used in each filter. The accuracy to which analogue filters can be matched is limited by the precision of the individual components. A bank of eight low-pass analogue filters matched in gain to 1% would be a major investment, yet the same system costs almost nothing to implement digitally.

Digital filters do have some disadvantages. They are very difficult to design and the subject is highly mathematical. Luckily, the design process can be automated and performed by a computer. They are also computationally intense. Both these problems have been solved by cheap computing power and good software, and using digital filters is now a breeze. In this section we are not going to go into details of how digital filters are designed, but will explore how they work and can be used to process data.

Using a digital filter starts with a specification. A very common requirement is for a low-pass filter, one that lets through frequencies up to a certain point and blocks higher ones. The filter is specified in exactly the same way as for an analogue filter (Chapter 3): the corner frequency f_c and the roll-off rate are selected. The classic filter types (Butterworth, Chebyshev, and Bessel) can all be implemented in digital form and are specified in the same way as their analogue counterparts. For example, we might specify a fourth-order Butterworth low-pass filter with a corner frequency of 10 Hz. Many programs are set up so that the values for the corner frequency and order can be typed in directly without any further calculations. The sampling rate and type of filter are also specified and the software does the rest.

The corner frequency of digital filters is sometimes not specified as a frequency in Hz but rather as a fraction of the Nyquist frequency. For example, if the corner frequency is 10 Hz and the sampling rate is 50 Hz (giving a Nyquist frequency of 25 Hz) then the relative corner frequency is $10/25 = 0.4$. The relative corner frequency can vary between 0 and 1 (there are no components present in the signal beyond the Nyquist frequency). Some programs specify the corner frequency as a fraction of the sampling rate, rather than the Nyquist frequency, and in this case the relative corner frequency can vary between 0 and 0.5.

There is no particular reason why one of the classic filter designs has to be used, and many digital filter design schemes work directly from the specifications to produce the filter that best fits them. So instead of a filter type and order, the roll-off rate, stopband attenuation and relative corner frequency are specified and the digital filter nearest to these specifications is constructed.

Digital filters are implemented in two stages. The design stage takes a set of specifications and produces a set of *filter coefficients*, a list of numbers that contain the key information needed to make the filter. This is the difficult part. The second part of the process takes these coefficients and uses them to perform the actual filtering, which is quite straightforward. Some data acquisition/analysis programs (e.g. SuperScope) have limited functions for filter design but you can calculate the filter coefficients with a separate program (e.g. Matlab) and then transfer them to SuperScope to perform the filtering itself. More advanced programs such as LabView do the whole process seamlessly, calculating the coefficients and then passing them to the filter routine.

As an example, the coefficients for a low-pass filter with a stopband attenuation of 60 dB and a cut-off frequency (relative to the Nyquist frequency) of 0.2 are listed in Table 5.1.

TABLE 5.1 COEFFICIENTS OF A LOW-PASS FIR DIGITAL FILTER WITH A STOPBAND ATTENUATION OF 60 DB AND A CUT-OFF FREQUENCY (RELATIVE TO THE NYQUIST FREQUENCY) OF 0.2

Coefficient number	Value
1	0.0002805487914
2	0.000335353225
3	2.074419054e-22
4	−0.0008684661522
5	−0.002054373737
6	−0.002885951648
7	−0.002430166051
8	−2.086519663e-21
9	0.004215193119
10	0.008755657363
11	0.01109807919

Coefficient number	Value
12	0.008608123511
13	1.97841275e-21
14	−0.01330610854
15	−0.02668023923
16	−0.03316062967
17	−0.020568730131
18	−3.018540759e-21
19	0.04294795938
20	0.09619524105
21	0.1481875037
22	0.1861093872
23	0.2
24	0.1861093872
25	0.1481875037
26	0.09619524105
27	0.04294795938
28	−3.018540759e-21
29	−0.02568730131
30	−0.03316062967
31	−0.02668023923
32	−0.01330610854
33	1.978412753e-21
34	0.008608123511
35	0.01109807919
36	0.008755657363
37	0.004215193119
38	−2.086519663e-21
39	−0.002430166051
40	−0.00288 5951648
41	−0.002054373737
42	−0.0008684661522
43	2.074419054e-22
44	0.000335353225
45	0.0002805487914

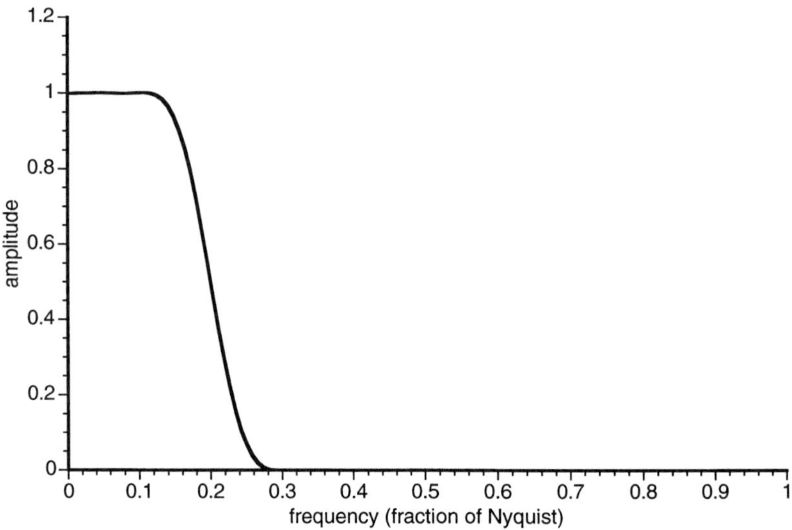

Figure 5.23 Frequency response of a low-pass FIR digital filter with forty-five coefficients. The design specifications were a cut-off frequency of 0.2 times the Nyquist frequency and a stopband attenuation of 60 dB.

The frequency response of this filter is shown in Figure 5.23. Note that

1. there are several coefficients (forty-five in all);
2. the coefficients are symmetrical about the middle value (0.2), which is necessary for a linear phase response;
3. there is an odd number of coefficients.

These properties are typical of this type of filter, which belongs to the class of *finite impulse response filters* (FIR). The other main class of digital filters is the *infinite impulse response filter* (IIR), and the key differences between the two types are shown in Table 5.2.

IIR filters used to be popular because the small number of coefficients made them fast. Now that the price of computing power has fallen, FIR filters are probably more widely used because their linear phase response distorts signals less. The phase response of a filter was discussed in Chapter 3: it describes how the delay introduced by the filter changes with frequency. A linear phase response means that the filter delays signals of all frequencies by the same amount. The signal is delayed by its passage through the filter but is not altered in shape except for the removal of the unwanted frequencies. A filter that does not have a linear phase response, on the other hand, will distort the signal to some extent.

TABLE 5.2 A SUMMARY OF THE MAIN DIFFERENCES BETWEEN FIR AND IIR DIGITAL FILTERS

Finite impulse response filters	Infinite impulse response filters
Relatively easy to design	Relatively difficult to design
Always stable	May go wild if not properly designed
Large number of coefficients for a given specification	Small number of coefficients
Usually linear phase response	Usually nonlinear phase response
Coefficient precision less important	Coefficient precision more important

Once the coefficients of a filter have been calculated, the filtering operation itself has to be implemented. For a given type of filter (FIR or IIR) the implementation is the same for all filters (low-pass, high-pass, etc.). To see how this is done, we will start with a popular but rather inferior type of FIR low-pass filter, the *moving-average filter*. This works by taking each point in the array and replacing it by the average of itself and a few points on either side of it. For a nine-point moving-average filter, each point is replaced by the average of itself and the four points on either side of it. The process is like having a small 'averager' which slides along the array, replacing each sample with the local average (Figure 5.24), and then moving along to the next sample. Each sample is not actually replaced straight away but stored for a little while until the averager has moved on so that the averager always works with the original array.

The way that the moving-average filter is implemented is typical of an FIR filter (Figure 5.25):

1. A section of the original array, the same length as the number of coefficients in the filter, is selected.
2. The coefficients are lined up alongside the array points.
3. Each array point is multiplied by the filter coefficient alongside it, and the result of all the multiplications is summed.
4. The sample in the centre of the selected section of the array is replaced by the sum (not immediately, it is stored until the filter has moved on).
5. The filter moves down the array one sample and the process is repeated.

There are several consequences of this activity. Firstly, filtering each point requires as many multiplications and sums as there are filter coefficients, which

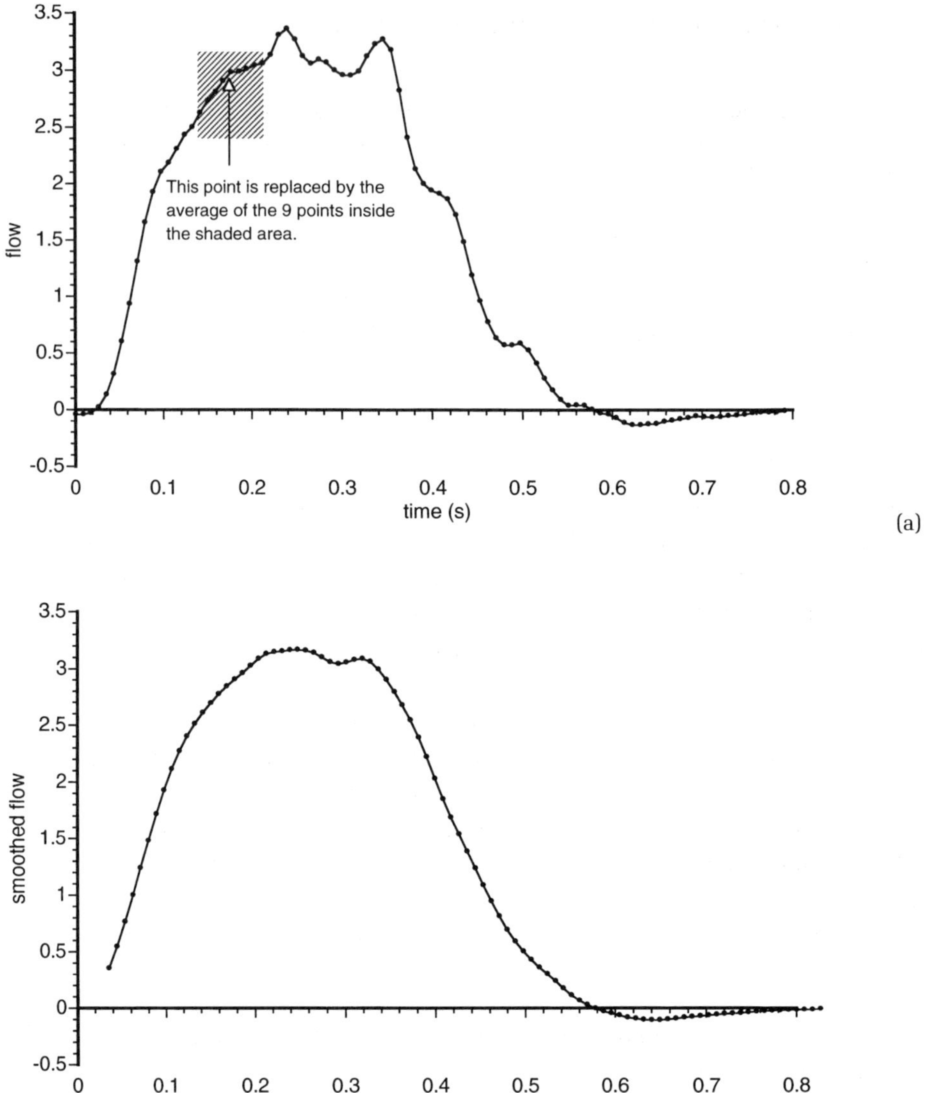

Figure 5.24 The effect of a nine-point moving average filter. (a) Raw data. The shaded area contains nine samples. The sample in the middle is replaced by the average of itself and the four points on either side of it. The averager then moves on one sample and the process is repeated. The averager always operates on the original samples and recently averaged samples are stored until the averager has moved on. (b) Filtered data. The signal is smoother and low-pass-filtered. The filter has also distorted the signal at the ends. The particular algorithm used to filter the data did not zero-pad the start of the data, so the first four points are missing because the mean value could not be calculated. On the other hand, the data were zero-padded at the end so the signal continues for longer than the original.

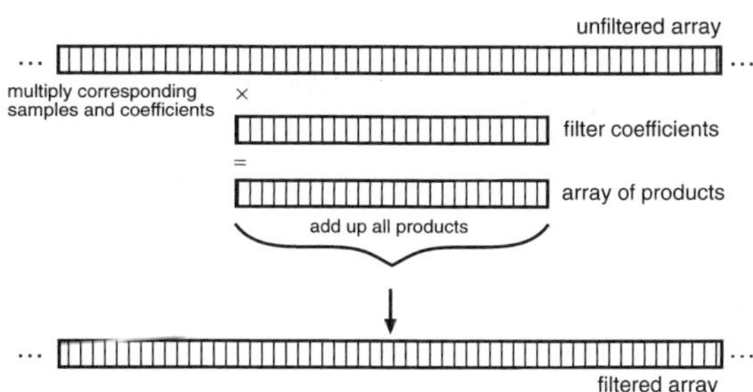

Figure 5.25 Mechanism of a FIR filter. The filter coefficients are lined up against the unfiltered array. Each array point is multiplied by the filter coefficient next to it, and all the resultant products are added together. The result is a filtered version of the sample next to the centre of the filter. The filter always operates on the original samples and newly filtered samples are stored until the filter has moved on.

is why short filters are faster. Secondly, the filter runs into problems at the start and end of the array because there are not enough samples with which to line up the filter. The usual solution to this problem is to extend the array with zeroes at each end.

The moving-average filter is just an FIR filter with each coefficient equivalent to 1/(number of points averaged). Thus a five-point moving-average filter is equivalent to an FIR with the coefficients

$$0.2, \quad 0.2, \quad 0.2, \quad 0.2, \quad 0.2$$

The moving-average filter is popular because it is very easy to design and incorporate into a program but it has a poor frequency response. Figure 5.26 shows the frequency response of a nine-point moving-average filter. The amplitude starts falling from 1 straight away, so the passband flatness is not good. The amplitude first becomes zero at a frequency of f_s/n, where f_s is the sampling rate and n is the number of points averaged. The amplitude response then rises and falls several times, producing a set of ripples or *sidelobes*. The corner frequency (the frequency at which the amplitude response has fallen to $1/\sqrt{2}$) is $0.443f_s/n$, where f_s is again the sampling rate and n is the number of points averaged.

For small numbers of filter coefficients (i.e. small values of n) the frequency response of a well-designed low-pass FIR (Figure 5.27) is not that much better than a moving-average filter and the sidelobes are about as large. One important difference, however, is that we can choose the corner frequency of the low-pass FIR whereas that of the moving-average filter is fixed by the filter length. As the number of filter coefficients is increased the differences between the two filters become more apparent. The shape of the frequency response of the moving-average filter does not change as the number of coefficients increases, it is just squashed to the left (Figure 5.26). So the sidelobes remain large and the passband

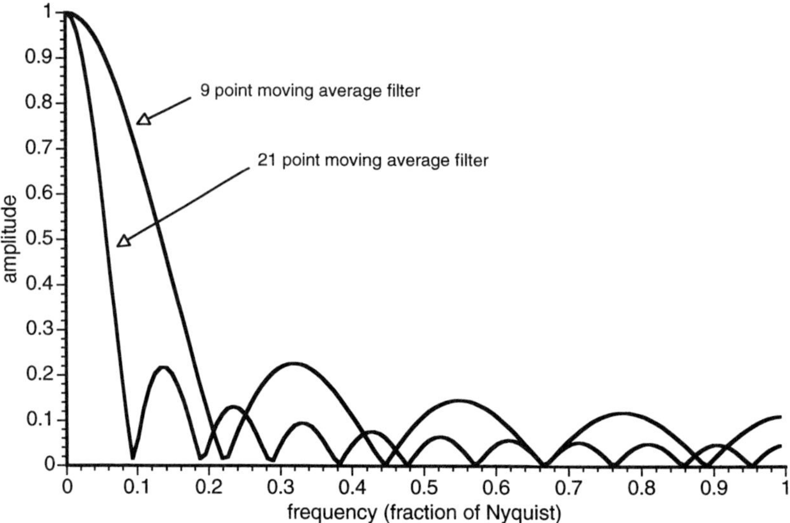

Figure 5.26 Frequency response of nine- and twenty-one-point moving-average filters. Note the large sidelobes and the fact that the filter amplitude response is linked to the number of points averaged. The moving-average filter's poor frequency response is not improved by averaging more points.

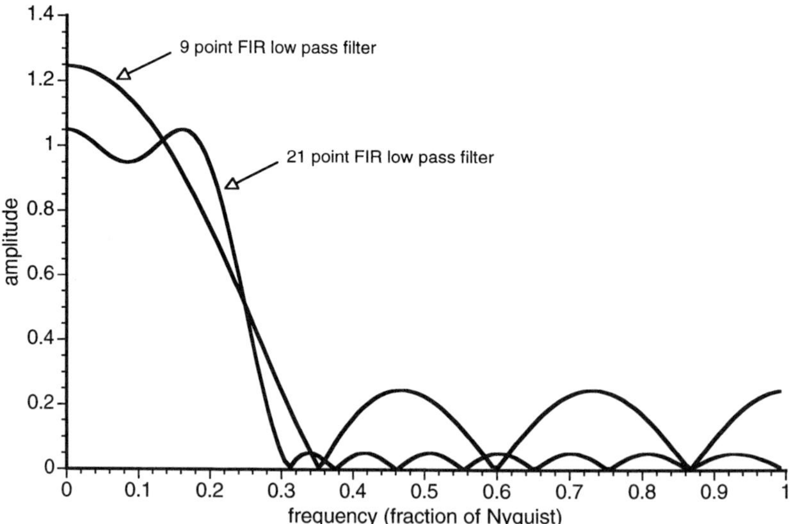

Figure 5.27 Frequency response of low-pass FIR filters with nine and twenty-one coefficients. The nine point filter is not much better than a nine-point moving-average filter except that the cut-off frequency can be varied. As the number of coefficients is increased the frequency response of the filter improves considerably whilst the cut-off frequency remains independent of the number of coefficients.

flatness is poor. The corner frequency becomes lower because it is set by the number of coefficients ($f_c = 0.443f_s/n$). In contrast, the more coefficients the FIR filter has, the better it becomes: the roll-off rate is steeper and the passband and stopband both become flatter (Figure 5.27). The corner frequency stays the same because it is independent of the number of coefficients.

IIR filters are implemented in a manner similar to FIR filters, but data points from both the original (unfiltered) array and the filtered array are multiplied by the filter coefficients. Because an IIR uses points from the filtered array (the output of the filter) as well as those from the unfiltered array (the input to the filter) an IIR has the potential to become unstable. When this happens the output of the filter either becomes larger and larger (until the computer cannot deal with the huge numbers any longer) or gradually slides closer and closer to zero. This should not happen with good filter design. However, IIRs can be marginally stable with a poor transient response. This means that they take a long time to settle after a sudden change in the input – the filter 'rings' (Figure 5.28).

Designing digital filters is a series of trade-offs. The most important one is the filter length (the number of coefficients) versus the 'quality' of the filter (how close its response is to that of an ideal filter). Figure 5.29 shows how the quality of a low-pass FIR improves as the number of coefficients increases. Another trade-off is passband/stopband flatness against roll-off rate. For a given filter length

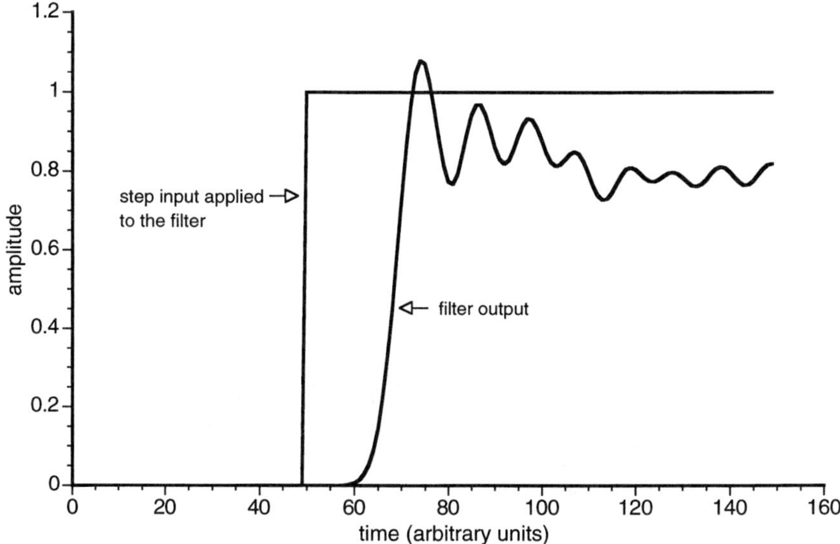

Figure 5.28 Ringing in a sharp cut-off (twelfth-order 2 dB Chebyshev) low-pass digital filter. A step input applied to the filter causes the output to oscillate for a while before settling down.

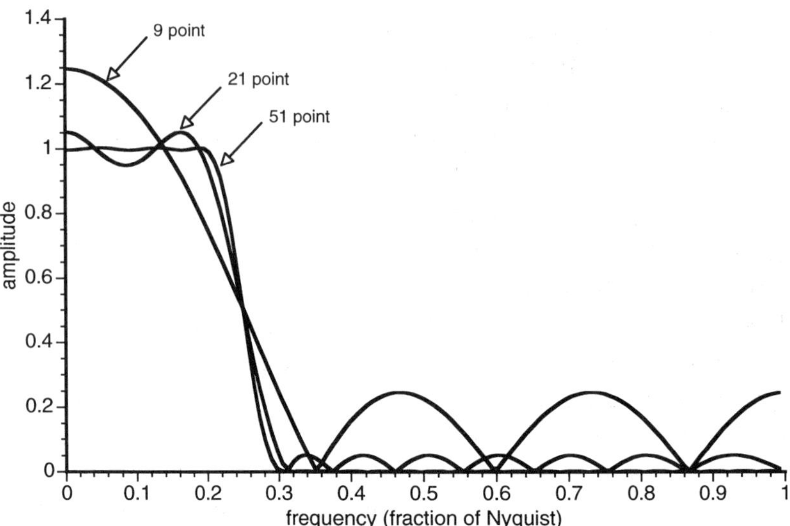

Figure 5.29 Frequency response of a low-pass FIR filter with nine, twenty-one, and fifty-one coefficients. The quality of the filter improves considerably as the number of coefficients increases.

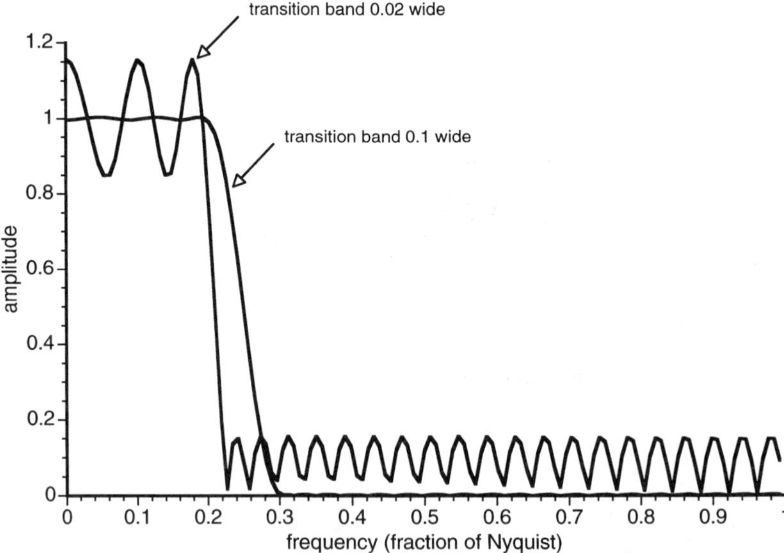

Figure 5.30 Frequency response of two fifty-one point FIR low-pass filters. For a fixed-length filter, there is a trade-off between passband/stopband flatness and roll-off rate. The sharp filter (transition band 0.02 units) has much more ripple compared to the softer filter (transition band 0.1 units).

the passband/stopband ripple increases as the roll-off rate increases. Figure 5.30 shows the frequency response of two fifty-one-point FIR filters, one with a wide transition band of 0.1 units and the other with a narrow transition band of 0.02 units (the frequency scale is dimensionless because it is specified in terms of f/f_n). The term *sharp* is sometimes used to describe filters: a sharp filter is one with a fast roll-off rate. You can also trade ripple in the passband for ripple in the stopband: the passband can be made flatter at the expense of more ripple in the stopband. The filters in Figures 5.27, 5.29, and 5.30 were designed with an algorithm (the Parks–McClellan algorithm) that gives equal ripple in the passband and stopband. A good filter design program will plot the filter's frequency response for you so that you can see how adjusting the number of coefficients affects the filter's performance.

Analogue filter design also involves trade-offs but they are different in nature. The main problem is the accuracy of the components. Given a particular filter design and set of specifications, the desired value of each capacitor and resistor can be calculated. Unfortunately, resistors and capacitors are only available in standard values so the one nearest to the calculated value has to be selected. Furthermore, the accuracy of each component's value is guaranteed only within a certain tolerance. Standard resistors are available to 1% tolerance. Capacitors are not quite as good and you have to use premium grade components (e.g. polypropylene) to get 2% tolerance. Once a set of standard component values closest to the desired ones has been chosen, the filter performance is recalculated to see how much it has been degraded by the errors in component values. Also, the calculation can be repeated many times, each time altering the values of the components slightly (but within their guaranteed tolerance) to see what the expected performance of a production run of filters would be. If too many filters will be out of specification, the design can be changed to one that is less sensitive to errors in component values but which uses more components, thus making it more expensive.

SPECTRAL ANALYSIS AND THE FAST FOURIER TRANSFORM

A very important tool in signal processing is the ability to change a signal from the *time domain* (a graph of the signal voltage against time) into the *frequency domain* (a graph of the amplitude and phase of the sine waves that make up the signal). The concepts behind this were discussed in Chapter 3. In this section the practical aspects of spectral analysis are explored.

The workhorse of digital spectral analysis is the fast Fourier transform (FFT), and almost all data acquisition/analysis programs will calculate the Fourier

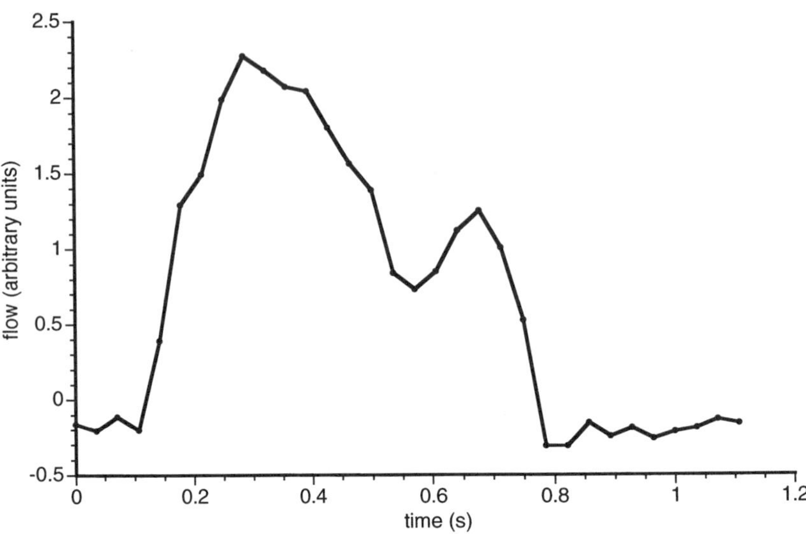

Figure 5.31 Respiratory flow rate during one inspiratory cycle, digitized at 28 Hz into thirty-two samples. The samples are shown by the small circles, the lines are for clarity only.

transform of a signal. The FFT is the name of a computer algorithm that was published by Cooley and Tukey in 1965. The algorithm is a fast way of calculating the discrete Fourier transform (DFT) of a signal and it is the DFT that does the transformation between the time and frequency domains. The details are available in most books on digital signal processing. In this book we will not go into any theory on how the FFT works, we will just use it as a signal-processing tool.

The FFT works on a block of data points. As an example, Figure 5.31 shows the respiratory flow rate during one inspiratory cycle that has been digitized into thirty-two samples at 28 Hz. The FFT of the waveform is plotted in Figures 5.32 and 5.33, and the original (time domain) and transformed (frequency domain) samples are listed in Table 5.3.

The first thing to note is that the FFT works on complex numbers.[3] It expects the time domain numbers to be complex, but this is impossible because they are samples of a real-world signal. So the imaginary part of each sample is zero (the data acquisition/analysis software usually takes care of this for you). The output of the FFT is the frequency-domain equivalent of the input, and this has both real and imaginary parts. Very commonly, the frequency domain is expressed in magnitude and phase format rather than real and imaginary and both forms have been included in Table 5.3. Complex numbers can be expressed in two

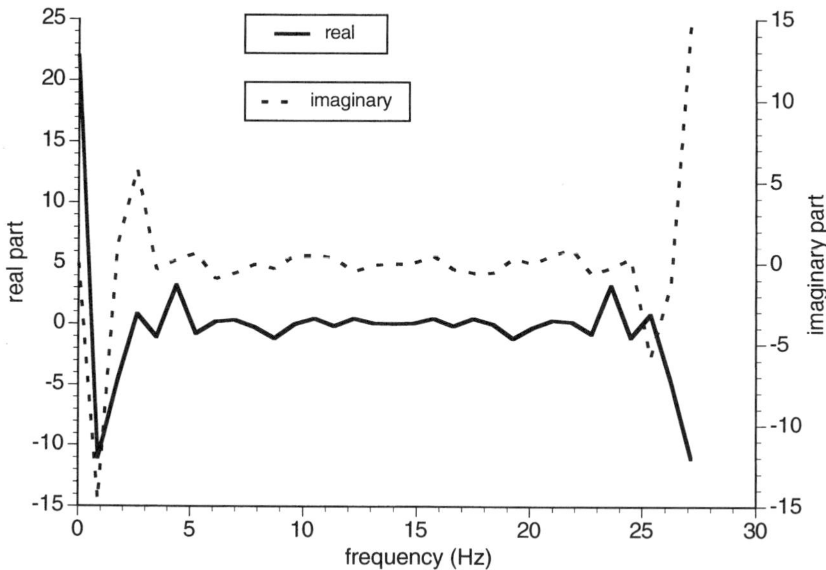

Figure 5.32 Spectrum of the signal in Figure 5.31 shown in real and imaginary form.

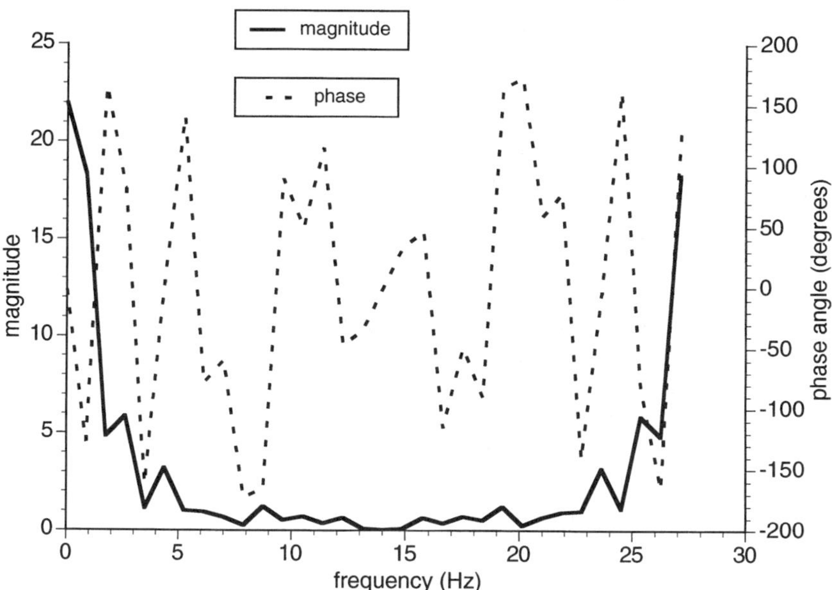

Figure 5.33 The same spectrum as Figure 5.32 but in magnitude and phase format. Note that the magnitude is reflected about the 14 Hz point. The phase angle is both reflected and inverted.

TABLE 5.3 THE ORIGINAL (TIME DOMAIN) WAVEFORM SHOWN IN FIGURE 5.31 AND ITS FFT (FREQUENCY DOMAIN) IN BOTH RECTANGULAR AND POLAR FORMS

Time domain		Frequency domain		Frequency domain	
Real	Imaginary	Real	Imaginary	Magnitude	Phase ($^\circ$)
-0.1643807	0	$2.1970728e+01$	$0.0000000e+00$	$2.1970728e+01$	$0.0000000e+00$
-0.2066356	0	$-1.1040449e+01$	$-1.4605201e+01$	$1.8308561e+01$	$-1.2708658e+02$
-0.1127359	0	$-4.6753049e+00$	$1.2452281e+00$	$4.8382919e+00$	$1.6508597e+02$
-0.2019406	0	$8.2372659e-01$	$5.7971990e+00$	$5.8554284e+00$	$8.1912952e+01$
0.3896275	0	$-1.0786999e+00$	$-3.5444253e-01$	$1.1354396e+00$	$-1.6181035e+02$
1.2910645	0	$3.1972758e+00$	$2.2285601e-01$	$3.2050331e+00$	$3.9871727e+00$
1.4929489	0	$-7.7555895e-01$	$6.3999037e-01$	$1.0055244e+00$	$1.4047059e+02$
1.9859223	0	$1.8942484e-01$	$-9.2988047e-01$	$9.4897811e-01$	$-7.8485888e+01$
2.2723165	0	$3.4273420e-01$	$-5.7748310e-01$	$6.7153069e-01$	$-5.9311017e+01$
2.1784167	0	$-2.6688594e-01$	$-3.3244227e-02$	$2.6894848e-01$	$-1.7289961e+02$
2.0704319	0	$-1.1943803e+00$	$-3.1624265e-01$	$1.2355378e+00$	$-1.6516979e+02$
2.0422621	0	$1.9408022e-03$	$5.2638499e-01$	$5.2638857e-01$	$8.9788749e+01$
1.8028178	0	$4.6835254e-01$	$5.4699407e-01$	$7.2010875e-01$	$4.9428882e+01$

1.5633737	0	$-1.6015615\mathrm{e}{-01}$	$3.3548727\mathrm{e}{-01}$	$3.7175489\mathrm{e}{-01}$	$1.1551905\mathrm{e}{+02}$
1.3896592	0	$4.4786414\mathrm{e}{-01}$	$-4.8098208\mathrm{e}{-01}$	$6.5721081\mathrm{e}{-01}$	$-4.7042011\mathrm{e}{+01}$
0.8403460	0	$8.1186501\mathrm{e}{-02}$	$-5.5346807\mathrm{e}{-02}$	$9.8257402\mathrm{e}{-02}$	$-3.4283203\mathrm{e}{+01}$
0.7323614	0	$4.6949700\mathrm{e}{-02}$	$0.0000000\mathrm{e}{+00}$	$4.6949700\mathrm{e}{-02}$	$0.0000000\mathrm{e}{+00}$
0.8497360	0	$8.1186501\mathrm{e}{-02}$	$5.5346807\mathrm{e}{-02}$	$9.8257402\mathrm{e}{-02}$	$3.4283203\mathrm{e}{+01}$
1.1220450	0	$4.4786414\mathrm{e}{-01}$	$4.8098208\mathrm{e}{-01}$	$6.5721081\mathrm{e}{-01}$	$4.7042011\mathrm{e}{+01}$
1.2535046	0	$-1.6015615\mathrm{e}{-01}$	$-3.3548727\mathrm{e}{-01}$	$3.7175489\mathrm{e}{-01}$	$-1.1551905\mathrm{e}{+02}$
1.0093654	0	$4.6835254\mathrm{e}{-01}$	$-5.4699407\mathrm{e}{-01}$	$7.2010875\mathrm{e}{-01}$	$-4.9428882\mathrm{e}{+01}$
0.5257820	0	$1.9408022\mathrm{e}{-03}$	$-5.2638499\mathrm{e}{-01}$	$5.2638857\mathrm{e}{-01}$	$-8.9788749\mathrm{e}{+01}$
-0.3099253	0	$-1.1943803\mathrm{e}{+00}$	$3.1624265\mathrm{e}{-01}$	$1.2355378\mathrm{e}{+00}$	$1.6516979\mathrm{e}{+02}$
-0.3099253	0	$-2.6688594\mathrm{e}{-01}$	$3.3244227\mathrm{e}{-02}$	$2.6894848\mathrm{e}{-01}$	$1.7289961\mathrm{e}{+02}$
-0.1549908	0	$3.4273420\mathrm{e}{-01}$	$5.7748310\mathrm{e}{-01}$	$6.7153069\mathrm{e}{-01}$	$5.9311017\mathrm{e}{+01}$
-0.2441955	0	$1.8942484\mathrm{e}{-01}$	$9.2988047\mathrm{e}{-01}$	$9.4897811\mathrm{e}{-01}$	$7.8485888\mathrm{e}{+01}$
-0.1878557	0	$-7.7555895\mathrm{e}{-01}$	$-6.3999037\mathrm{e}{-01}$	$1.0055244\mathrm{e}{+00}$	$-1.4047059\mathrm{e}{+02}$
-0.2582804	0	$3.1972758\mathrm{e}{+00}$	$-2.2285601\mathrm{e}{-01}$	$3.2050331\mathrm{e}{+00}$	$-3.9871727\mathrm{e}{+00}$
-0.2113306	0	$-1.0786999\mathrm{e}{+00}$	$3.5444253\mathrm{e}{-01}$	$1.1354396\mathrm{e}{+00}$	$1.6181035\mathrm{e}{+02}$
-0.1878557	0	$8.2372659\mathrm{e}{-01}$	$-5.7971990\mathrm{e}{+00}$	$5.8554284\mathrm{e}{+00}$	$-8.1912952\mathrm{e}{+01}$
-0.1315158	0	$-4.6753049\mathrm{e}{+00}$	$-1.2452281\mathrm{e}{+00}$	$4.8382919\mathrm{e}{+00}$	$-1.6508597\mathrm{e}{+02}$
-0.1596857	0	$-1.1050559\mathrm{e}{+01}$	$1.4605201\mathrm{e}{+01}$	$1.8308561\mathrm{e}{+01}$	$1.2708658\mathrm{e}{+02}$

forms: *rectangular* (also called *Cartesian*) form (real and imaginary) or *polar* form (magnitude and phase). It is easy to swap between the two forms:

$$\text{real part} = \text{magnitude} \times \cos(\text{phase})$$

$$\text{imaginary part} = \text{magnitude} \times \sin(\text{phase})$$

and, conversely,

$$\text{magnitude} = \sqrt{\text{real}^2 + \text{imaginary}^2}$$

$$\text{phase} = \tan^{-1}\left(\frac{\text{imaginary}}{\text{real}}\right)$$

The units of the output of the FFT are the same as those of the time domain signal. In this example, if the flow is in $l\,s^{-1}$ then the magnitude and real and imaginary parts of the spectrum are also in $l\,s^{-1}$. The units of phase angle are either radians or degrees, with radians being the more common.

The next thing to note is that there are the same number of points in the frequency domain as in the time domain. Only the first half of the frequency domain is of any use. In the second half of the frequency domain the real part is just a mirror image of the first half and the imaginary part is an inverted mirror image (see Figure 5.32); thus the second half of the frequency domain does not contain any new information. The 'mirror image' property arises because all the imaginary parts of the time domain are zero. Many data acquisition/analysis packages do not display the second half of the frequency domain because there is no point.

The frequency domain is interpreted as follows. Remember that any signal can be made up from a number of sine waves of different frequency and phase angle (Chapter 3), and that the FFT breaks the signal up into these individual sine waves. The *second* number in the output of the FFT is the amount present in the original signal of a sine wave whose frequency is such that *one* cycle will just fit into the total time taken to record the block of samples. In this example, thirty-two samples were recorded at 28 Hz so the total time taken to record those samples is 32/28 = 1.14 s. If one cycle of a sine wave just fits into 1.14 s then its frequency is 1/1.14 = 0.88 Hz. So the second number in the frequency domain is the amount of a 0.88 Hz sine wave that is present in the original signal. The magnitude is the amplitude of the sine wave, and the phase is its phase angle relative to the start of the sampled signal.

The *third* number in the output of the FFT is the amount present in the original signal of a sine wave whose frequency is such that *two* cycles will just fit into the total time taken to record the block of samples. The frequency of this sine wave is 1.75 Hz, and its amplitude and phase are specified by the magnitude and phase angle. The fourth number in the output of the FFT is the amount present in the original signal of a sine wave whose frequency (2.63 Hz) is such that three cycles will just fit into the total time taken to record the block of samples, and so on. In general, the sine wave frequency is given by

$$\text{frequency} = (\text{bin number} - 1) \times f_s/N$$

where f_s is the sampling rate (28 Hz in this example) and N is the total number of samples (32 in this example). The numbers that make up the output of the FFT are often referred to as *bins*. Bin number 1 is the first value in the frequency domain, bin number 2 is the second value, and so on.

This pattern continues until we reach bin 16, the last one before the half-way mark. This bin contains the amount of a 13.13 Hz sine wave present in the original signal. What about bin 17? This would contain the amount of a 14.0 Hz signal. But we sampled at 28 Hz, so the Nyquist frequency is 14 Hz and all the components in the original signal are below this frequency. There is no new information in the frequency domain above the half-way point. This is true no matter how many data points were collected and no matter what the sampling rate. All the frequency-domain information is contained in the first half of the FFT output.

Finally, back to the first bin. Using the equation above, this contains the amount of a sine wave of 0 Hz present in the original signal. A signal of 0 Hz is a DC signal, so what the first bin contains is the amount of DC present in the time domain signal.[4] It is easier to think of this as the mean value of the time-domain signal, scaled by some factor. The phase angle and the imaginary part are always zero.

Some important practical points about using the FFT for frequency analysis are emerging:

1. The width of a bin is 1/(total time taken to record the block of samples). The fundamental resolution of the FFT output is one bin. The frequency resolution of the FFT is therefore determined solely by the recording time. If it takes 4 s to record the block of samples the frequency resolution is 0.25 Hz.

   If there are N samples in the block, recorded at a sampling rate of f_s (in Hz), then the total time taken to record them is N/f_s seconds. Thus the frequency resolution is also equal to f_s/N Hz.

2. The frequency range of the output of the FFT is from 0 Hz (DC) up to one bin below the Nyquist frequency.
3. Each FFT bin contains a complex number. This represents the amount present in the original signal of a sine wave whose frequency is given by

$$\text{frequency} = (\text{bin number} - 1) \times f_s/N$$

where f_s is the sampling rate and N is the total number of samples.

There are some restrictions on what N (the number of samples in the block) can be. The FFT is most efficient when N is an integer power of 2 (2, 4, 8, 16, 32, etc.). Typical values for N are 128 to 512 samples. The most common and most efficient version of the FFT is called a *radix-two* FFT, and it requires that N is a power of 2. Most programs use a radix-two FFT and force you to use a value for N that is a power of 2. Matlab is a bit more intelligent: it looks at N and decides which version of the FFT to use, so you can use other values of N if you wish but it will take longer to do the calculations. The most inefficient value of N is a prime number, which forces the program to use the unmodified DFT.

Often it happens that the number of samples to be transformed is not a power of 2. If the FFT program will not accept this there are two standard solutions to the problem. The first is to truncate the signal, that is, throw samples away, until the next lowest power of 2 is reached. The second is to pad the sample at each end with zeros until the next highest power of 2 is reached. There is no obvious choice as to which is better. If you have a long recording of a signal that is repetitive, so that the recording has captured many cycles of the signal, then truncating the signal will not make much difference to its spectrum (as calculated by the FFT). For example, the spectrum of a blood pressure recording will be similar whether the recording spans twenty beats or thirty. On the other hand, if the recording has captured a single event such as a muscle twitch, it is better to pad the recording with zeros because truncating it will alter the shape of the signal and thus its spectrum.

A minor difference between algorithms is shown by taking the transform of a steady 2 V signal (i.e. every sample in the input array is 2.0). This is a pure DC signal so the first frequency domain bin contains a wholly real number (the phase angle and imaginary part are zero) and all the other bins are zero. Using Matlab or Excel to perform the FFT the first bin contains 16. However, using SuperScope to perform the calculation the first bin contains 4. This is because the algorithm used in SuperScope scales all the frequency components by $2/N$. In general, the exact values in the FFT bins are not that important. We nearly always want to compare

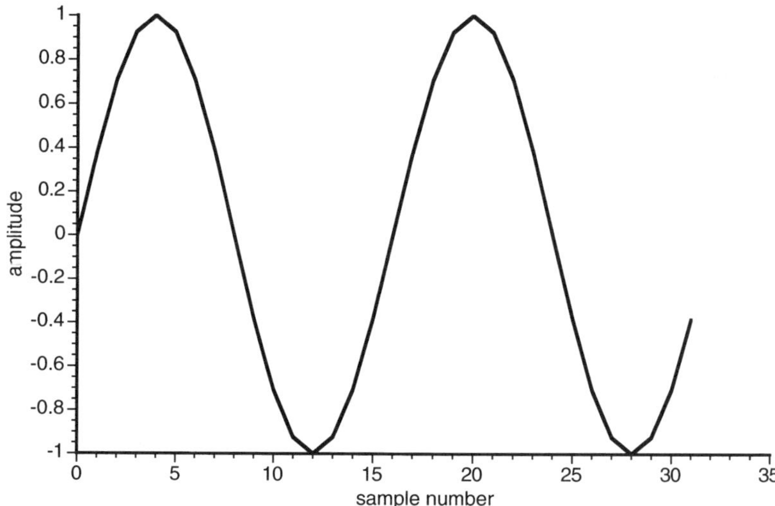

Figure 5.34 A sine wave with amplitude 1.0 and frequency such that two cycles are spread exactly over thirty-two samples.

the contents of one bin with another, and thus we are interested in ratios of bin amplitudes, on which scaling of the results has no effect.

Figure 5.34 shows a digitized sine wave with a frequency such that two cycles exactly fill the input array of thirty-two samples. The amplitude of the sine wave is $1.0.$[5] The FFT of this signal contains zero in each bin except for bin 3, which contains $0 - 16j$. This means that the only frequency component present in the input array is a sine wave of a frequency such that two cycles fit the input array. The fact that the number in bin 3 is purely imaginary means that the phase angle is $-90°$, indicating that the wave is a sine wave. If the input wave is a cosine wave then bin 3 contains $16 + 0j$. These values can easily be converted to magnitude and phase format, using the equations given earlier. The magnitude of bin 3 for both the sine wave and the cosine wave is $\sqrt{0^2 + 16^2} = 16$. The phase angle for the sine wave is $\tan^{-1}(-16/0)$. Although $-16/0$ is undefined, we know that in this context it represents the tangent of $-90°$. For the cosine wave the phase angle is $\tan^{-1}(0/16) = 0$. The fact that a cosine wave is associated with a phase angle of 0, rather than a sine wave, is easier to understand if the polar form of the output is written out in full. The complex number $a + jb$ is expressed in polar form as

$$a + jb = re^{j\theta} = r(\cos\theta + j\sin\theta)$$

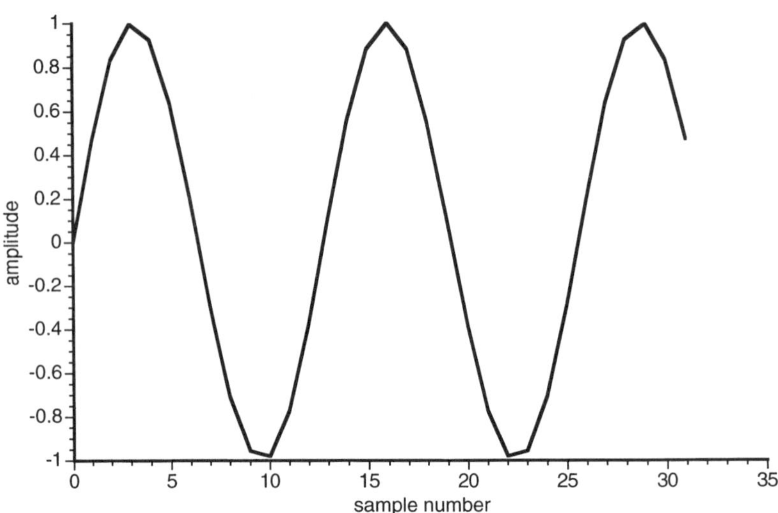

Figure 5.35 A sine wave with amplitude 1.0 and 2.5 cycles spread over thirty-two samples.

where a is the real part, b the imaginary part, r is the magnitude, θ is the phase angle and $j^2 = -1$. The value in bin 3 for the sine wave is $0 - 16j$, or in polar form $16\{\cos(-90°) + j\sin(-90°)\}$.

The magnitude of bin 3 is 16, yet the amplitude of the sine wave is 1.0. This is because the standard FFT scales the amplitude of signals by $N/2$, where N is the number of samples. In this case the sine wave was digitized into thirty-two samples so the amplitude of the sine wave (1.0) was multiplied by 32/2. The FFT of long sequences of samples (e.g. 256 or 512 samples) therefore has large amplitudes but this is of little practical importance because, as mentioned above, we are almost always interested in ratios of magnitudes, not their absolute values.

Now fill the input array with 2.5 cycles of a sine wave (Figure 5.35) and calculate the spectrum (Figure 5.36).[6] The output is quite different. All the bins contain frequency components. Bins 3 and 4 have the greatest magnitude and it tapers off gradually on each side. This effect is called *leakage*. It arises because the FFT assumes that the signal is continuous, and that the input array is just a snapshot of this continuous signal. The FFT effectively looks at a continuous stream of input arrays stacked end to end, not just a single one. So if an exact number of cycles fit into the input array the continuous signal looks like Figure 5.37 – a pure sine wave. But if 2.5 cycles fit into the array then the continuous signal looks like Figure 5.38. Clearly, this is not a pure sine wave and the sudden jump between the end of one array and the start of another adds extra frequency components that

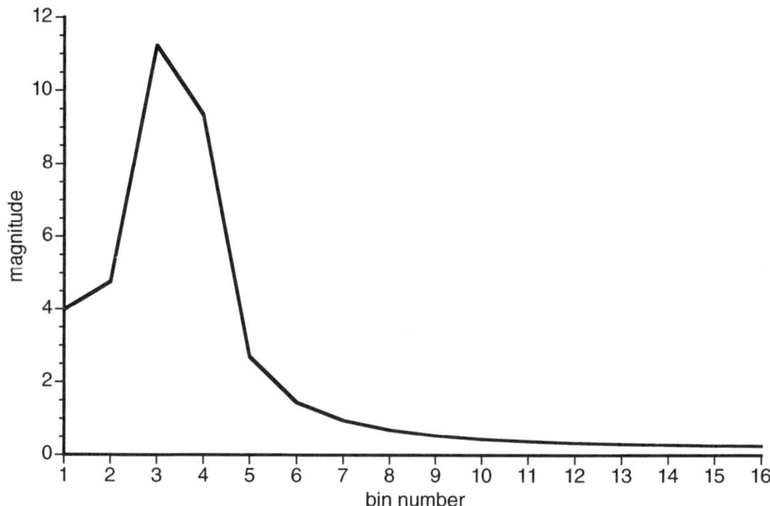

Figure 5.36 Magnitude of the spectrum of the signal in Figure 5.35. Every bin contains some signal component due to leakage.

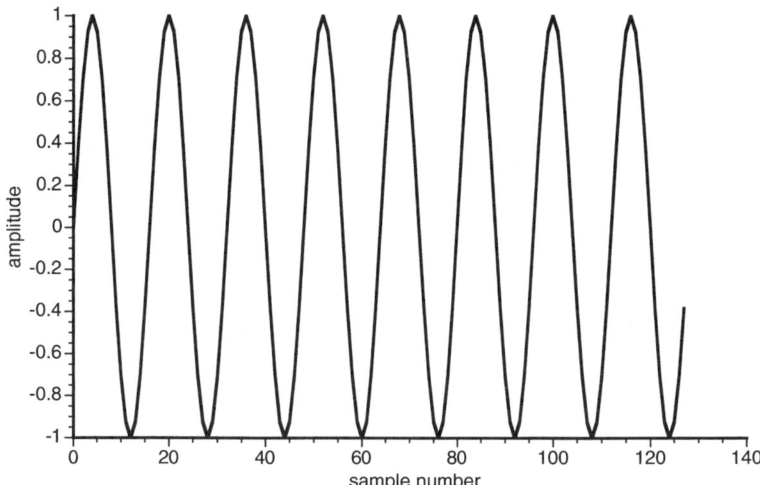

Figure 5.37 Multiple copies of the signal in Figure 5.34 stacked end to end. The result is a pure sine wave.

appear in the output. The average value is also not zero so there is a component in the first bin.

Leakage affects all signals and is undesirable because it 'smears' the contents of one bin into those on either side. If we were interested in a second small oscillation present in the original signal that gave a component in bin 6, for example, the

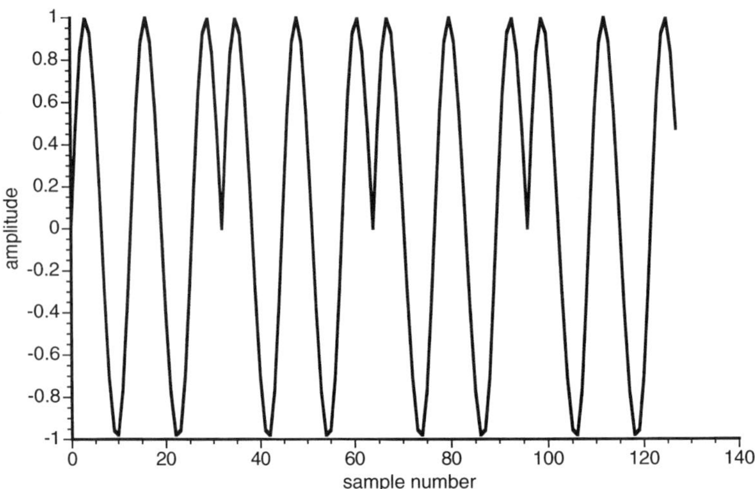

Figure 5.38 Multiple copies of the signal in Figure 5.35 stacked end to end. The discontinuities at the junctions contain high- and low-frequency components and the mean value is not zero.

results would be distorted by leakage from the lower-frequency sine wave. Another point to note is that the bulk of the amplitude of the sine wave is shared between bins 3 and 4, being the two bins closest in frequency to the sine wave.

Leakage cannot be eliminated entirely when using the FFT, but its effects are usually reduced by a technique called 'windowing'. The array of samples is multiplied by a weighting function before being transformed. The weighting function has the effect of reducing the amount of signal at the edges of the input array so that the discontinuities between arrays are not as large. When transformed, there is less leakage of the spectrum into other bins (Figure 5.39). However, like everything in digital signal processing, there is a trade-off. A sine wave that would fill just one bin when transformed now fills two or even three after windowing, so that spectral resolution (the ability to distinguish between two frequency components close together) is reduced.

There are many different types of windows. In general, they trade leakage for spectral resolution: the windows that give the lowest leakage (e.g. the Kaiser window) also have the worst spectral resolution. A window that is widely used is the Hamming window, and its shape is shown in Figure 5.40.[7] All windows consist of a set of symmetrical coefficients whose value ranges from 1 in the centre to near zero at the edges. There are as many coefficients as there are samples in the input array. The window and the samples are 'lined up' next to each other and corresponding samples and coefficients are multiplied together. The effect is to

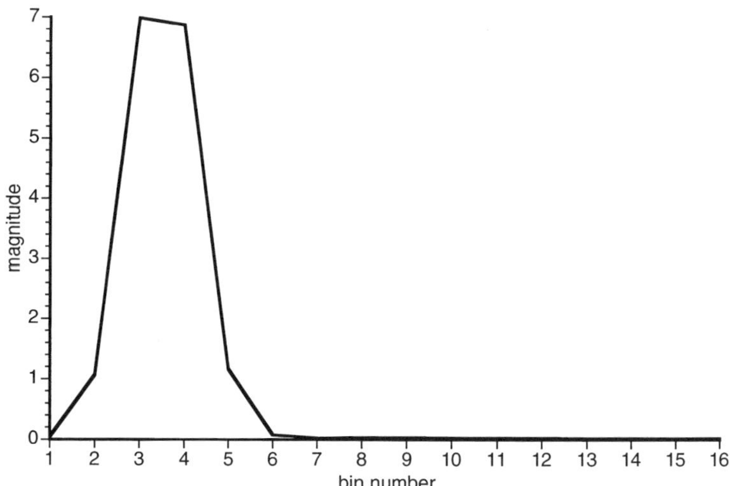

Figure 5.39 The same spectrum as Figure 5.36 but with a Hamming window applied to the data before the FFT. There is less leakage but the spectral resolution is not as good. The DC component (bin 1) is also smaller.

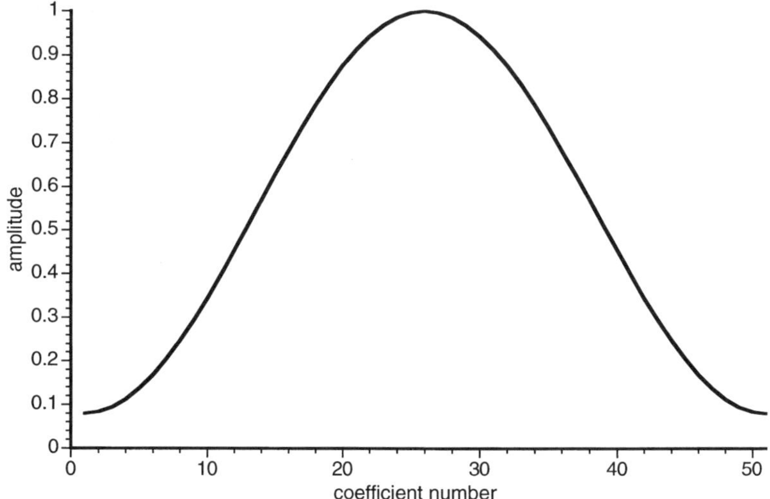

Figure 5.40 A fifty-one point Hamming window.

shape the input samples by the 'envelope' of the window. The windowed array is then converted into the frequency domain using the FFT.

It is generally accepted that data should be windowed before using the FFT unless you have good reason not to, and all data acquisition/analysis packages

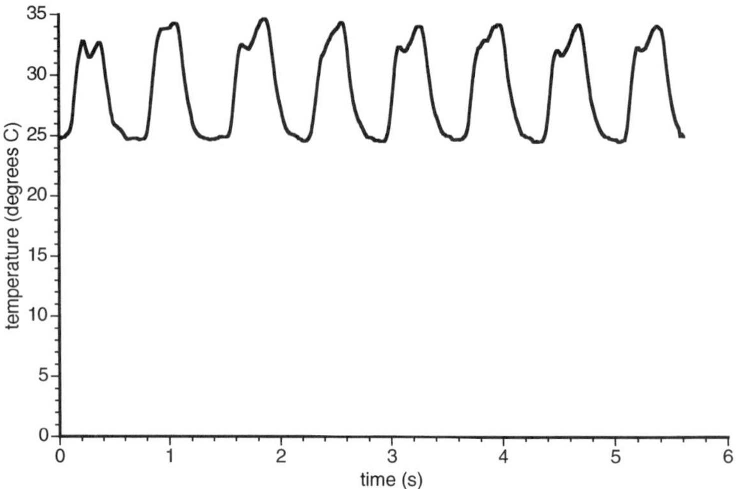

Figure 5.41 Recording of the air temperature at the nostril of a horse during exercise, made with a fast responding thermistor. The ambient temperature is 24.8°C and the expired air is at about 35°C. The mean temperature is 28.8°C.

have windowing functions built in. The choice of which window to use is less important, the Hamming window being a popular all-rounder. The more complex Kaiser window allows the user to decide how much spectral resolution he/she wishes to trade for leakage reduction by varying the parameter β. The larger the value of β the lower the leakage and the poorer the spectral resolution. For details of how to set β, see Rabiner and Gold, p. 101.

If there is a DC component present in the input array (i.e. the mean value of the signal is not zero) then this appears in the first bin in the output of the FFT. This will be smeared to some extent into the adjacent bins (especially if the data are windowed), making it difficult to get accurate measurements of the low-frequency components. It takes only a relatively small amount of DC to produce a large component in the first bin, which is often called the *DC spike*. In spectral analysis we are not usually interested in the size of the DC component and so it is removed before performing the FFT. This is quite easy: the mean of the input array is calculated and the mean value is then subtracted from every point in the input array. This operation is called *DC spike removal* and is performed before windowing. Figure 5.41 is a recording of the air temperature at the nostril of an exercising horse. The temperature varies from ambient (24.8°C) to around 35°C with each breath. Figure 5.42 is the magnitude of the spectrum of the temperature (the trace was truncated from 633 to 512 samples and a Hamming window applied before transforming) from 0 to 15 Hz. There is a large DC spike present that leaks into the

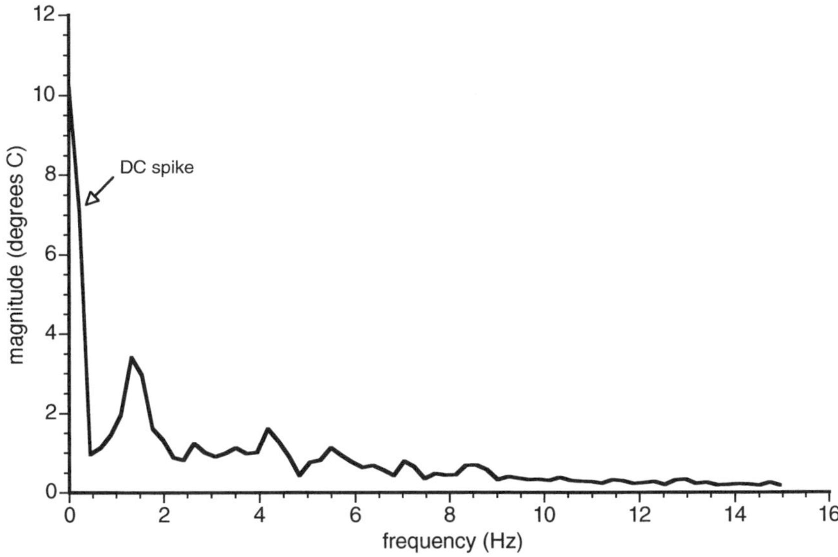

Figure 5.42 Magnitude of the spectrum of the signal in Figure 5.41. There is a large DC spike that has been smeared into the first couple of bins by the Hamming window.

first few bins. Figure 5.43 is the magnitude of the spectrum of the same recording but with the average value of the temperature (28.8°C) subtracted. The DC spike has gone and the largest peak is now at 1.32 Hz (79.2 breaths/min), the respiratory rate.

The phase information in the spectra of biological signals is often noisy and difficult to interpret (see Figure 5.33, for example). Very frequently we are only interested in 'how much' signal is present at a certain frequency, and thus spectral parameters that do not include phase information are useful. The simplest is the magnitude of the spectrum. This is always positive and is a useful indicator of the amount of signal present at a particular frequency. A related parameter is the *power spectral density*, often just called the *power spectrum*. The power spectral density is the amount of energy in the signal in a given frequency range. It is easy to calculate from the spectrum:

$$\text{power spectral density} = \frac{1}{N}(\text{magnitude})^2$$

where N is the number of samples used to calculate the spectrum, and magnitude is just the magnitude of the spectrum. Another way of expressing the power spectrum is

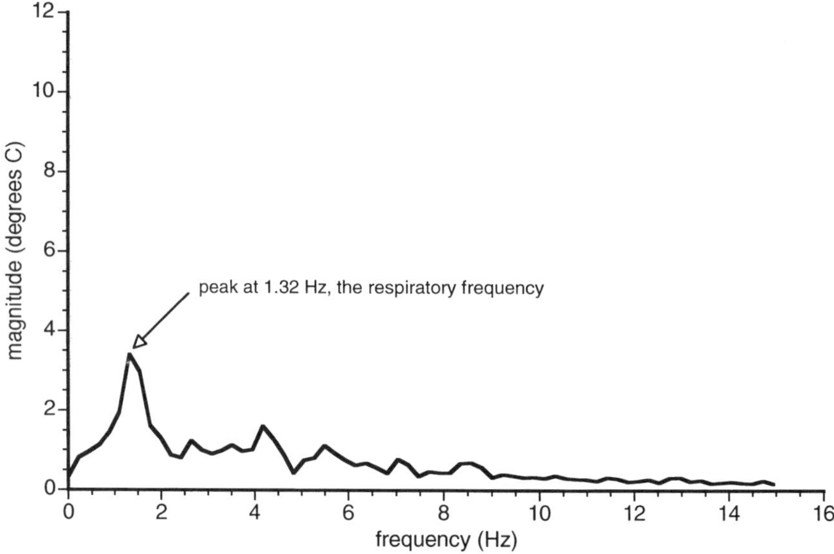

Figure 5.43 The same spectrum as Figure 5.42 but with the average value of the temperature subtracted from the data before windowing. The DC spike has gone and it is easier to pick out the peak at the respiratory rate of 1.32 Hz.

$$\text{power spectral density} = \frac{1}{N}(\text{real}^2 + \text{imaginary}^2)$$

where real and imaginary are the real and imaginary parts of the spectrum. The units of the power spectrum are $\text{units}^2\,\text{Hz}^{-1}$, that is, if the units of the sampled data are °C then the units of the power spectrum are $°\text{C}^2\,\text{Hz}^{-1}$.

The power spectrum is important because, as mentioned with RMS values, power is a good way of comparing signals that have very different waveforms. Thus if we want to compare the relative amounts of two signals at a certain frequency, comparing their power spectra is a good way to go about it. Although the power spectral density is a measure of the energy in a given frequency band, the FFT gives us the spectrum already divided into bins that have a fixed width and thus

$$\frac{1}{N}(\text{real}_i^2 + \text{imaginary}_i^2)$$

is a measure of the amount of signal power in bin number i, where real_i and imaginary_i are the real and imaginary parts of bin i. Note that if you are calculating ratios of power at different frequencies in the same power spectrum, or comparing ratios of spectra of arrays with the same number of samples, then the value of N

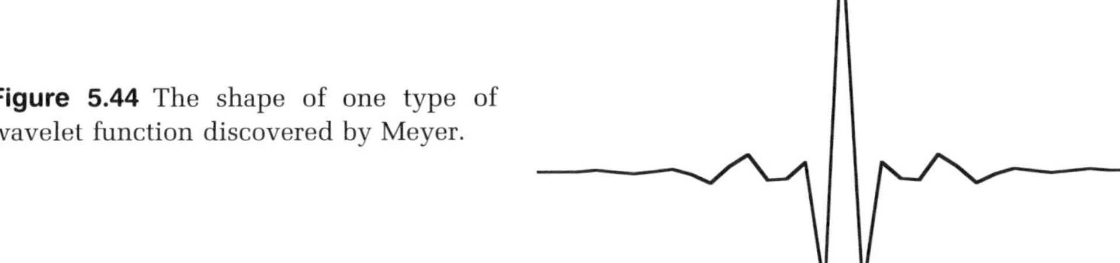

Figure 5.44 The shape of one type of wavelet function discovered by Meyer.

disappears from the calculations. For example, if you wish to compare the amount of signal power in bin 5 compared to that in bin 3, the calculation is simply

$$\frac{\text{power in bin 5}}{\text{power in bin 3}} = \frac{\text{real}_5^2 + \text{imaginary}_5^2}{\text{real}_3^2 + \text{imaginary}_3^2}$$

where $\text{real}_5 =$ real part of bin 5, and so on. In dB the ratio is

$$\frac{\text{power in bin 5}}{\text{power in bin 3}} = 10 \log_{10}\left(\frac{\text{real}_5^2 + \text{imaginary}_5^2}{\text{real}_3^2 + \text{imaginary}_3^2}\right) \text{dB}$$

Before leaving spectral analysis, other transforms should be mentioned. None of these is used as much as the FFT. The fast Hartley transform (Bracewell 1984) provides the same information as the FFT but is generally quicker to calculate because it accepts purely real input information, unlike the FFT. Like the FFT, the fast Hartley transform (FHT) is an algorithm for efficient calculation of the discrete Hartley transform (Hartley 1942). The FHT is gradually making its way into commercial programs and LabView has a built-in FHT. Other methods for calculating the spectrum of a signal have advantages over the FFT such as freedom from leakage and resolution that is independent of the recording length, but the algorithms are more involved and have to be programmed from scratch because they are not supported by signal-processing packages. The comprehensive review by Kay and Marple (1981) covers many of these methods. Finally, wavelet analysis moves away from sine waves and represents signals as collections of small 'blips' (Figure 5.44). As the 'frequency' increases the 'blips' have the same overall shape but become narrower and taller. Wavelet transforms can describe the shape of some signals with just a few coefficients and are becoming very important in image compression because images compressed and expanded using wavelets show less distortion than other methods.

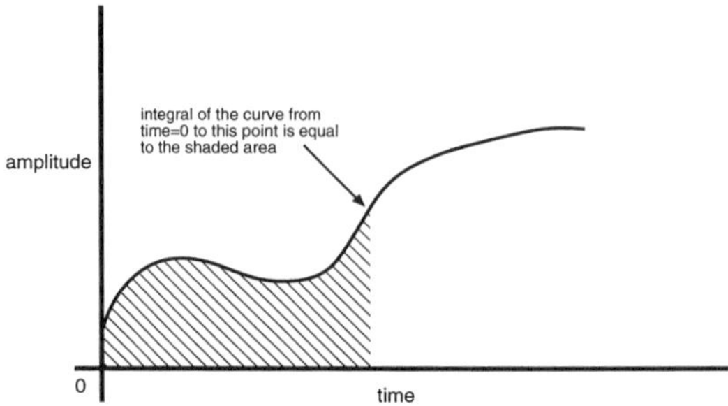

Figure 5.45 The integral of a curve is equal to the area underneath it (hatched area).

INTEGRATION

Integration of signals (with respect to time) is a common task. In respiratory work the usual way of quantifying air movement is by measuring flow with a pneumo-tachograph. The flow is then integrated to give volume. Motion analysis often uses accelerometers, and the signal from these can be integrated once to give velocity and twice to give distance. Indicator dilution techniques for measuring cardiac output involve measurement of the area under the indicator–time curve, which again requires integration.

Numerical methods for integration are well-known mathematical techniques that are readily transferred to digital signal processing. When a signal is integrated it is the same process as calculating the area underneath the voltage–time graph (Figure 5.45). The simplest integration method uses the trapezoidal rule. A straight line is drawn between each point and the area between the two points is equal to $(V_1 + V_2)/2\delta t$, where V_1 and V_2 are the amplitudes of the two points and δt is the time between the two points, that is, the sampling interval ($\delta t = 1/f_s$, where f_s is the sampling rate in Hz). The trapezoidal rule is easily extended to a string of samples:

$$\text{integral of } V \text{ (volt–seconds)} = \delta t(V_1/2 + V_2 + V_3 + \cdots + V_{n-1} + V_n/2)$$

where V_1, V_2 are the values of successive samples of which there are n in all (Figure 5.46). Very often, the factor of 1/2 applied to the first and last samples is ignored and the integral is then δt(sum of all the samples).

More accurate integration methods are available but are not used that often. The simpler ones all involve applying weighting coefficients to the samples. For example, Simpson's rule for integration is

$$\text{integral of } V \text{ (volt–seconds)} = \delta t(V_1/3 + 4V_2/3 + 2V_3/3 + 4V_4/3 + \cdots$$
$$+ 4V_{n-1}/3 + V_n/3)$$

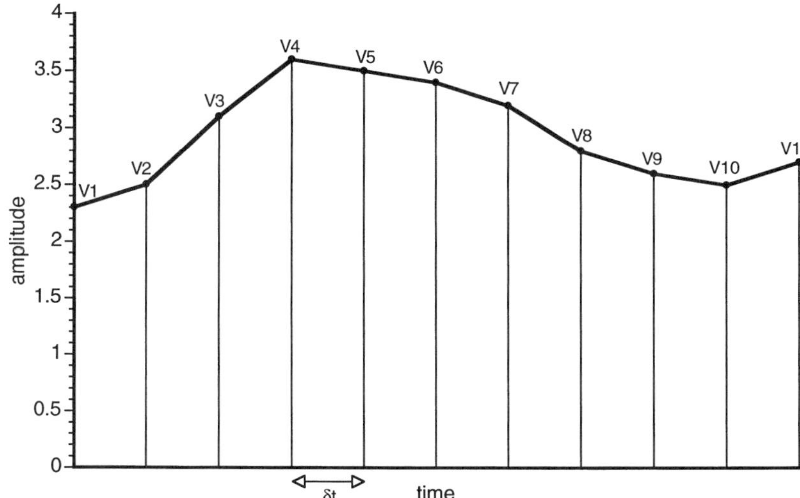

Figure 5.46 Eleven points of a signal sampled δt seconds apart. Joining up the samples and the time axis forms ten trapezii. The integral of the original continuous signal is approximated by the sum of the areas of the ten trapezii.

Numerical Recipes in C (Chapter 4) goes into the errors associated with integration methods in some detail.

Before digital computers were readily available, signals were integrated by analogue means with electronic circuits. One of the biggest bugbears of these circuits was their tendency to 'drift'. Even a slight DC voltage present caused the output of the integrator to change over time. With digital integration the drift has been eliminated from the integrator, but it is worth remembering that a small DC offset will still cause a significant error if the integration time is long. In general, integrations should not be performed over long periods of time without 'resetting' the integrator. A slightly different cause of drift is found when integrating respiratory flow to obtain tidal volume. The integrator must be reset to zero at the end of each inspiration and expiration. If this is not done the integrator output will rise or fall steadily, because inspired and expired tidal volumes are not the same.

Integrators smooth out signals and reduce noise. This is because integrators act as low pass filters. The frequency response of an integrator is shown in Figure 5.47. The amplitude is proportional to 1/frequency, so an integrator reduces higher frequencies more than the low ones. This can be useful if the data you are working with have some high-frequency noise, because if the final result can be obtained by integration it will be relatively insensitive to the noise. In contrast, an analysis

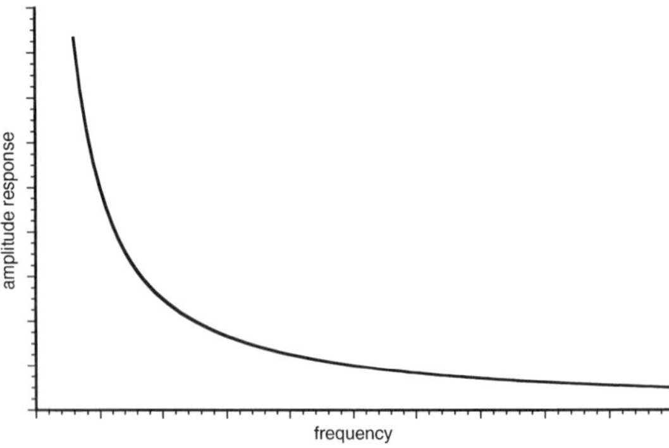

Figure 5.47 The frequency response of an integrator. It acts as a low-pass filter. The response at DC is infinite and the integral of a DC voltage does not have a fixed value but a steadily rising voltage.

method that relies on identifying some point on a waveform (e.g. picking out the local maxima) will be very sensitive to noise.

DIFFERENTIATION

Like integration, there are both simple and more involved ways of finding the derivative of a signal. The simple method uses the approximation $dV/dt \approx \delta V/\delta t$ (Figure 5.48). If V_1 and V_2 are the amplitudes of two successive points and δt is the time between the two points, that is, the sampling interval

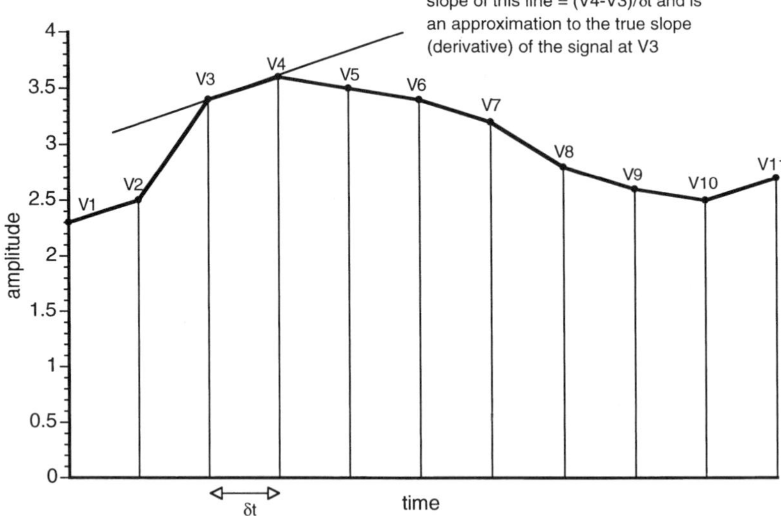

Figure 5.48 Eleven points of a signal sampled δt seconds apart. The slope of a line joining adjacent points is approximately equal to the differential of the original signal at that point.

$(\delta_t = 1/f_s$, where f_s is the sampling rate in Hz), then $\delta V = V_2 - V_1$ and $\delta V/\delta t = (V_2 - V_1)/\delta t$. This is then extended to the whole array by calculating the difference between successive samples:

$$\text{derivative of } V \text{ (V/s)} = \frac{1}{\delta t}(V_2 - V_1, V_3 - V_2, \ldots, V_n - V_{n-1})$$

This results in a derivative array that is one shorter than the original array. There is also a shift in time of half a sample, because the value calculated for the derivative really belongs to a point half-way between V_1 and V_2. One way to get round the shift is to calculate the derivative using the points on either side, that is, the derivative at position 2 is given by $(1/\delta t)(V_3 - V_1)$, but this is less accurate. Normally the time shift caused by differentiation is not important unless you are measuring time differences between differentiated and undifferentiated signals, and even then a shift of half a sample is unlikely to be important. If it is, the sampling rate probably needs to be increased.

More accurate methods of calculating dV/dt are based on fitting a curve through a few adjacent points on the original array. The equation of the curve is known so its derivative can be calculated analytically. The method avoids the time shift and rejects some noise (see below) but is obviously much more difficult to program. SuperScope uses a fourth-order polynomial to approximate the curve, and fits the polynomial over five points. *Numerical Recipes in C* (sections 5.7 and 14.8) discusses different types of filters and gives results for the Savitzky–Golay filter.

One problem with signal differentiation is noise. The frequency response of a differentiator is shown in Figure 5.49. It acts as a high-pass filter because the amplitude is proportional to the frequency, and thus high-frequency noise is amplified. This is quite a nuisance in practice because there is always some

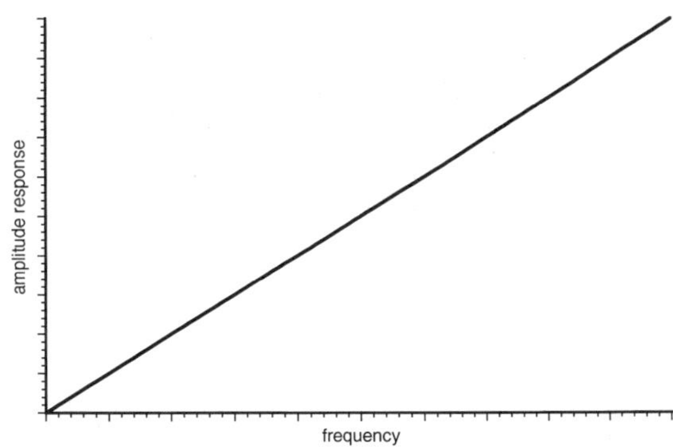

Figure 5.49 The frequency response of a differentiator. It acts as a high-pass filter, accentuating any high-frequency noise.

high frequency noise present with the signal. If further derivatives are taken (d^2V/dt^2, etc.) the problem becomes even worse. It is important, therefore, to apply a low-pass filter to data before taking the derivative.

CORRELATION

Correlation is the process of finding out 'how much' of one signal there is in another. It is often used to look for a pulse that we know is somewhere in a signal but the noise makes it difficult to say just where. Figure 5.50 shows a recording which consists of noise and a small pulse. Looking at the recording, one can see that the pulse exists, but it is difficult to locate with any precision. To locate the pulse using correlation, a short signal consisting of just the expected shape of the pulse (it does not have to be that accurate) is constructed and used as a 'template'. The template is 'held up' against the recording to see how well it matches. The actual value that describes how well the template matches the recording (the correlation between the two) is calculated by multiplying adjacent samples in the smooth pulse and the long array, and then finding the mean of the products by integrating them (often just the sum is used) and dividing by the number of products. The template is then moved along one sample and the process repeated

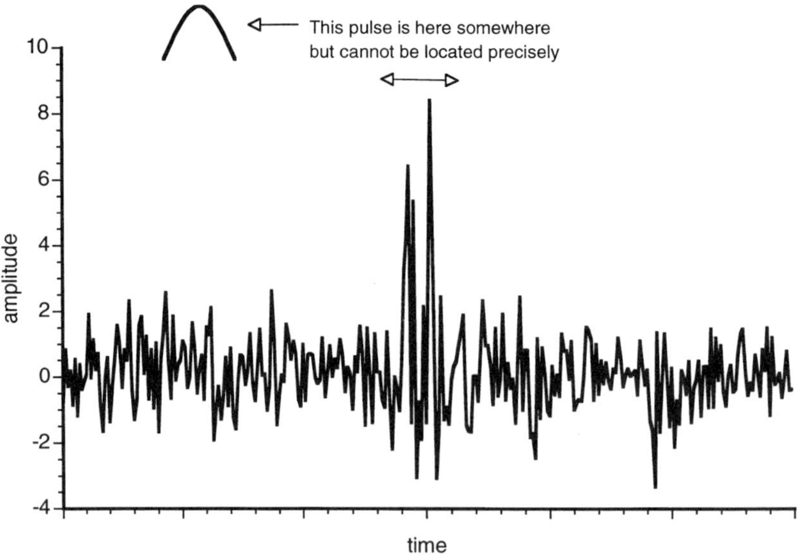

Figure 5.50 A recording of a pulse (inset) buried in noise. The position of the pulse is known only approximately.

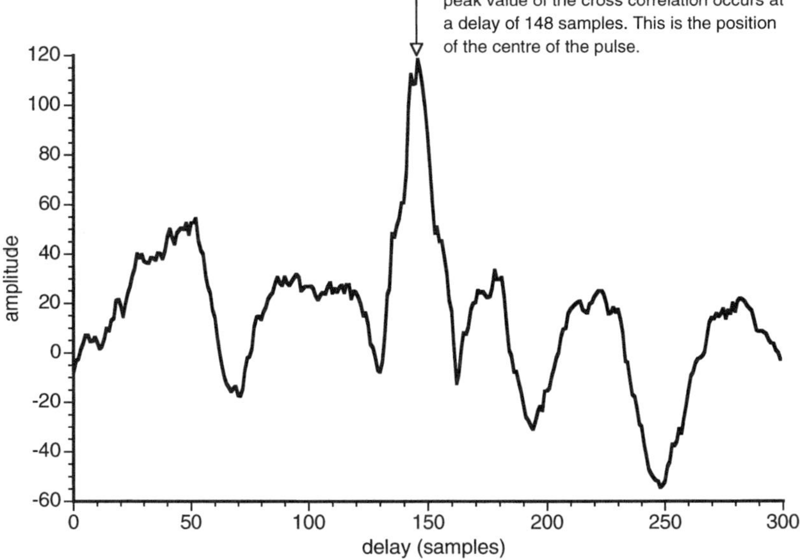

Figure 5.51 The result of cross-correlating a copy of the pulse with the noisy recording. The peak in the amplitude at a delay of 148 samples indicates the place where the best match is found. The true position of the pulse was at 150 samples.

until the end of the array is reached. The exact algorithms used to perform correlation vary somewhat: the last stage, dividing by the number of products, is just a scaling step and is sometimes not performed.

The answer obtained has two components: the position of the template along the array (the *delay, τ*) and the size of the correlation (the amplitude). This is plotted in Figure 5.51. The peak in the correlation corresponds to the best match between the template and the array, and the delay is how far from the start of the array this occurred. In this example the true position of the peak of the pulse was at a delay of 150 samples and the cross correlation yielded a value of 148, an error of 0.7%. Thus by using correlation we can calculate the location and quality (how well it matches the template) of the pulse in the presence of a fair degree of noise. Correlation has the power to extract signals that are almost buried in noise because the integration process filters it out.

In this example two different signals were correlated against each other. This is called *cross-correlation*. You can also correlate a signal against a copy of itself: *autocorrelation*. Although the latter may seem less useful, it is an important intermediate step in some signal-processing techniques.

There is clearly a problem at the end of the signal array, because the template has nothing to line up against. This problem is solved by extending the array with zeros. In the extreme case where both the signal array and the template are of the same length (as occurs in autocorrelation), one array has to be padded with zeros until its length is two times the recorded length less one sample.

The process of correlation has many similarities to other signal-processing tasks such as FIR filtering. In fact, it is possible to perform correlations by converting the signals to the frequency domain, manipulating them and converting the result back into the time domain. For large arrays this is much more efficient than the straightforward approach, which is sometimes called the *brute-force* method, because of the computational efficiency of the FFT.

CHAOS AND NONLINEAR ANALYSIS

Some signals that appear to be random noise may contain more information than is apparent. One physiological example that has been extensively studied is *heart rate variability*. If the time between heart beats (the interbeat or *R–R interval*) is measured accurately for many beats it is not constant, even if the subject sits still so that changes due to exercise are eliminated (Figure 5.52). The smoothed spectrum of the R–R interval (with the DC component removed and a Hamming window applied) is shown in Figure 5.53. There is a prominent peak at the

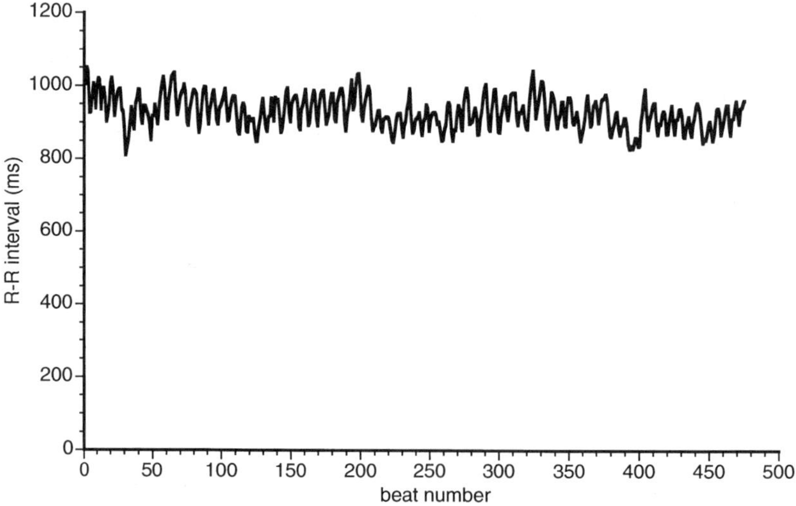

Figure 5.52 The time between successive heartbeats for a subject sitting quietly. The heart rate is not constant and the small fluctuations in rate hold important information about the sympathetic nervous system.

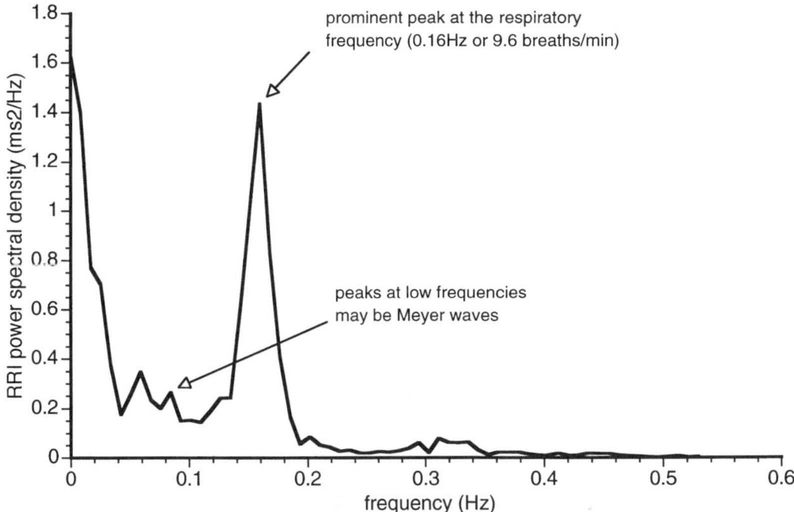

Figure 5.53 Power spectrum of the data in Figure 5.52. The prominent peak at the respiratory rate is due to respiratory sinus arrythmia.

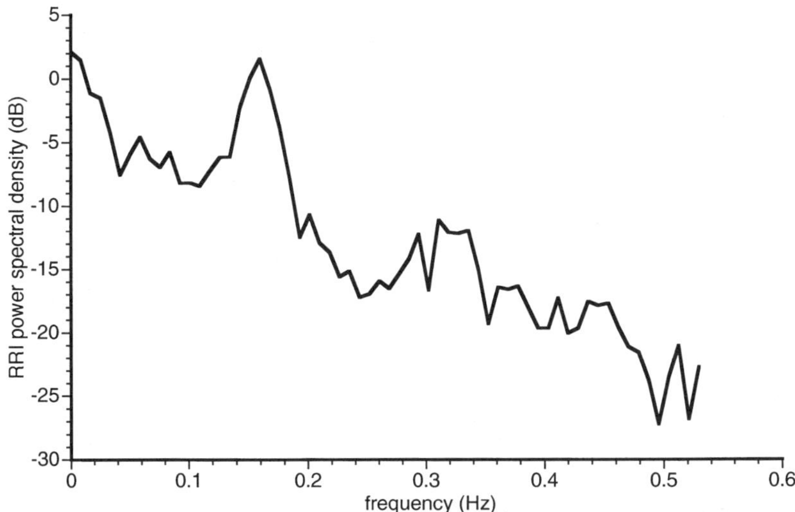

Figure 5.54 The same spectrum plotted in dB, showing the broad range of frequencies involved in heart rate variability.

respiratory rate (respiratory sinus arrythmia) and possibly some Meyer waves at 0.6 Hz and 0.8 Hz.[8] Figure 5.54 shows the same data but with the power spectrum in dB and Figure 5.55 shows the data plotted with both axes in logarithmic units.[9] How well the log–log plot approaches a straight line is of interest because a

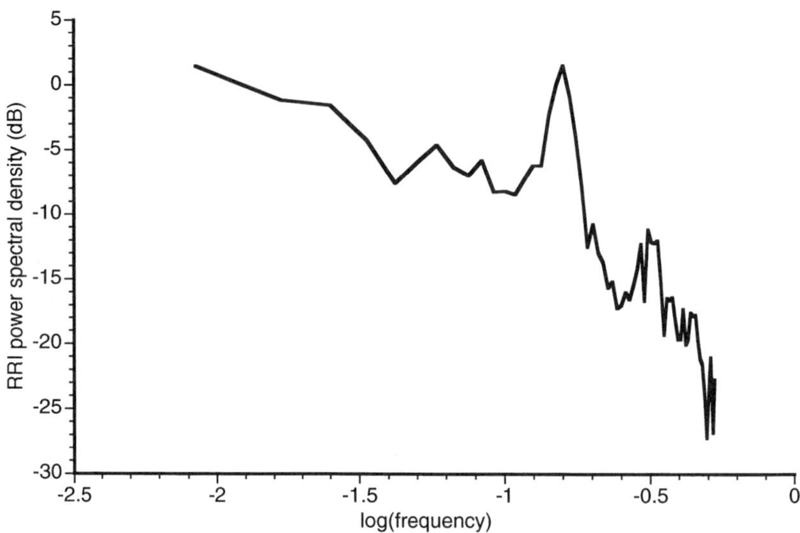

Figure 5.55 The same power spectrum as Figure 5.53 but plotted in log–log units. The result is close to a $1/f$ spectrum over most of the frequency range.

straight line on a log–log plot implies that the relationship between the data sets is of the form

$$\text{RRI power spectrum} = a \times \text{frequency}^{b}$$

and the slope of the line on the log–log plot is b. A negative value of b means that the power spectrum is of the form

$$\text{RRI power spectrum} \propto \frac{1}{\text{frequency}^{\beta}}$$

and such spectra are called $1/f$ *spectra*. They have attracted interest because $1/f$ spectra are characteristic of chaotic processes. Further analysis of such data involves removing the harmonic components (Yamamoto and Hughson 1991) and calculating nonlinear parameters such as the correlation distance and the fractal dimension. There is evidence that the normal chaotic nature of the heart rate is lost in cardiac disease, and analysis of heart rate (which can be obtained noninvasively) may yield early predictors of cardiac disease. Nonlinear signal analysis is a specialized area and there is little commercial software available, although programs developed by researchers are freely available.[10]

REMOVING 50/60 Hz INTERFERENCE

Interference at 50/60 Hz from mains wiring is a constant problem unless you use battery-powered equipment in the middle of a field far from any power lines. The 50/60 Hz noise is usually present at mV levels so if the signal under investigation is large (a few V) then 50/60 Hz noise will probably not be a problem. As the signal becomes smaller, 50/60 Hz noise becomes more significant and if you are measuring signals in the μV range (EEGs, for example) it can be very difficult to reduce the 50/60 Hz noise to acceptable levels. The best approach is to stop the noise before it gets into the amplifier. The noise is picked up by the 'front end' of the signal-processing system, that is the preamplifier and amplifier. The noise enters the preamplifier and amplifier stages by two processes: capacitative coupling and magnetic coupling. Capacitative coupling is a problem with high impedance probes (pH and other ion-selective probes, intracellular electrodes, etc.) but is reduced by using screened or shielded cables.[11] Electrophysiology work is particularly demanding in terms of noise rejection. Glass microelectrodes have a resistance of several megohms and to avoid distorting the signal the preamplifiers have to have an input impedance many times higher than this, that is several gigohms ($10^9 \Omega$), which means that capacitative coupling of noise is a major problem. At the same time the system has to be sensitive enough to record currents in the picoamp (pA) range with a frequency response extending to many kHz. The electrode preamplifier is the critical component that has to meet these difficult specifications and its construction is very specialized. Selected low-noise components, resistors in the range of 50 GΩ and Teflon connectors are all crammed into a small box that mounts right next to the cell bath so that the distance between the cell and the amplifier (which is unshielded) is as short as possible.

Magnetic coupling is a problem with long wires and is reduced not by screening but by changing the wiring layout (Figure 5.56). A number of tricks are used to reduce mains interference, and in practice one tries them all until the interference is reduced to acceptable levels:

1. Move all power cables as far away as possible from electrodes and signal cables.
2. Use screened cable for all signals. All circuitry should be inside metal boxes. If you need to construct amplifiers, filters, etc. do not put them in a plastic box because this will not provide any shielding. In electrophysiology work it can be very difficult to get rid of all the interference and sometimes people have to resort to wrapping parts of the apparatus in aluminium foil and grounding them in order to provide a temporary shield.

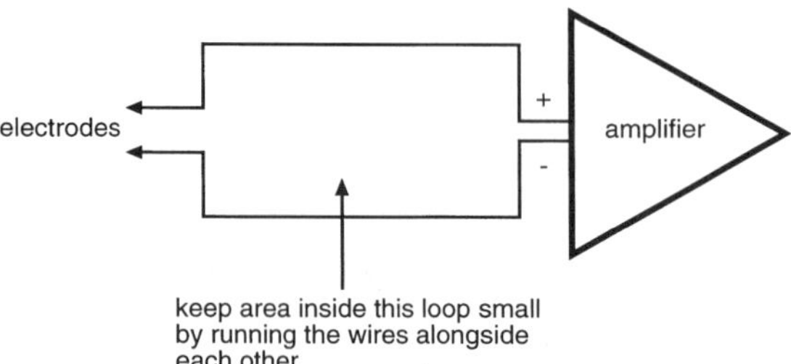

Figure 5.56 To avoid magnetic pick-up of mains interference by the input leads, minimize the area enclosed by the two wires.

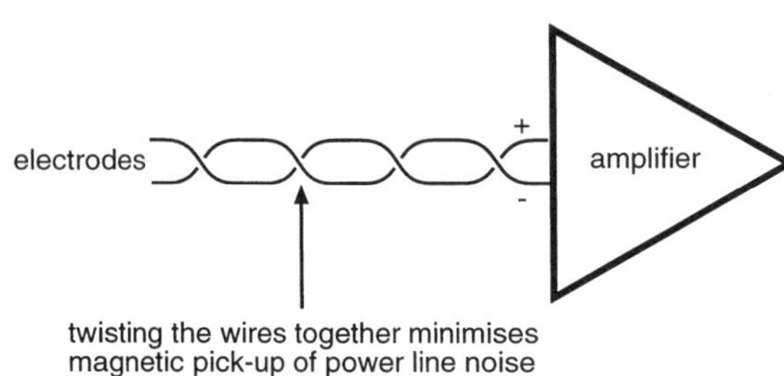

Figure 5.57 Twisting the input leads together is a good way of reducing magnetic pick-up of mains interference.

3. Do not form loops in the electrode wires (Figure 5.56) by keeping the signal cables close to each other. Even better, twist the wires together (Figure 5.57).
4. Reduce electrode impedance as much as possible.
5. Make sure electrode impedances are equal to each other. Use the same type of electrode at each measurement point.
6. Use amplifiers with differential inputs (Chapter 2).
7. If the amplifier is connected to a patient (ECG, EEG measurements, etc.) make sure the ground connection from the amplifier to the patient is good. Do not connect any other ground to the patient (this can be dangerous as well) and make sure the patient is on an insulated surface.
8. Use good-quality amplifiers. The type of circuitry used can make a big difference in the amount of noise that is rejected, and newer circuits are much better than old designs. In the past the only way that some measurements could be made with the equipment available at the time was to work inside a metal cage, and whole rooms were built inside metal boxes. You still see these 'screened rooms' in older institutions, but with modern amplifiers they are not needed for most work.
9. Pay careful attention to equipment grounding. For safety reasons almost any piece of equipment that plugs into the wall has a grounded (three-pin) plug on it. The ground pin on the plug is connected to the metal parts of the equipment and, in particular, the case if it is made of metal. Any input or

output sockets on the equipment that are shielded will probably have their shield connected to the ground pin as well.

One problem with this system occurs when several pieces of equipment are joined together in a string by signal cables, for example, a preamplifier, amplifier, and chart recorder. Each of these pieces of equipment also plugs into the wall and the continuous ground connnection between the boxes creates a *ground loop*. The loop starts at the wall, goes down the ground wire in the power cable to the first piece of equipment, through the shielding in the cables and returns to the wall via the power cable of the last piece of equipment. This loop picks up interference just like a loop in the electrode wires, plus extra interference if there is a fault current flowing in the ground wiring inside the wall. One thing that you can do is use a *star grounding* arrangement. As far as possible, run all the equipment from one outlet (using a power bar) or at least from outlets that are close to each other, not on opposite sides of the room. There is little risk of overloading the outlet because signal amplifiers do not consume much power. You may get some indication that you have a ground loop because the interference will stop when you break the loop by disconnecting two pieces of equipment, such as an amplifier and a filter unit. The interference will return if you reconnect just the shielding of the cable by touching the outer part of the cable plug to the socket on the equipment.

10. In some cases, particularly in neurophysiology recordings, the apparatus around the electrodes (microscope, micromanipulator, etc.) needs to be grounded to prevent capacitative pick-up of interference. The ground connection used for this should be the signal ground from the preamplifier and not the power ground in the wall, if this is possible. Use a star-earthing pattern so that the grounding wires from each piece of equipment come back to the same point on the preamplifier, and avoid 'daisy-chaining' ground connections from one piece of apparatus to another.

In spite of all these tricks, it can be difficult to get rid of all the 50/60 Hz noise from the signal and what is left has to be filtered out. One common way is to use a 'notch' filter. This is a filter with a frequency response that falls to zero over a narrow range of frequencies. The filter can be made in either hardware (filtering the signal before it is digitized) or in software. Another useful trick is to sample at exactly the same frequency as the interference. Surprisingly, this rejects all signals at that frequency. Figure 5.58 shows why: the same point on the waveform is sampled each time so the contribution from the noise signal is zero.

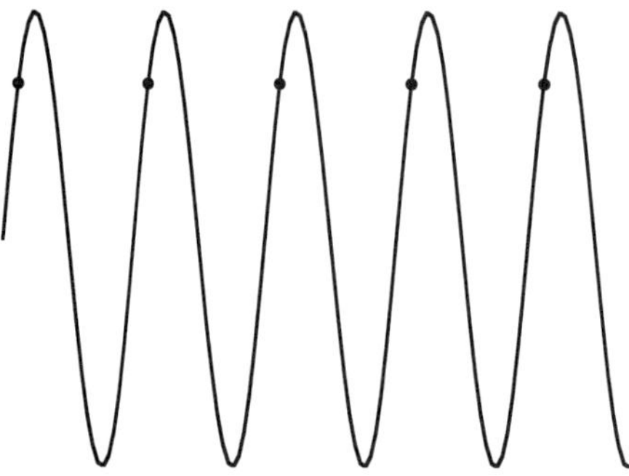

Figure 5.58 Sampling at the same frequency as a signal rejects that signal completely because the signal effectively does not change between samples.

The effectiveness of filtering is limited by the fact that the power line frequency is not exactly 50/60 Hz but varies by about ±10% during the day. One can make the filters 'broader' (i.e. reject a wider range of frequencies) but this also rejects more of the signal. An alternative is to measure the frequency of the power line noise at the time that the signal was recorded and adjust the filter or sampling rate accordingly. This is known as *tracking* the noise. Tracking filters are usually digital: it is possible, but difficult and expensive, to make a hardware-based automatic tracking filter. Because tracking filters are able to alter their characteristics depending on some property of the signal (in this case the frequency of the noise component) they are known as *adaptive* filters.

A different approach to 50/60 Hz noise removal is used in the 'Hum Bug', made by Quest Scientific (*www.quest-sci.com*). The Hum Bug is a digital device that identifies the 50/60 Hz noise present in the signal and tracks it. However, instead of using a notch filter to remove the noise the Hum Bug constructs a replica of the noise and subtracts it from the signal. In this way the phase shifts associated with filters are avoided. Also, the Hum Bug removes not just the fundamental frequency (50/60 Hz) but all the associated harmonics up to 4 kHz.

TWO- AND THREE-DIMENSIONAL SIGNAL ANALYSIS

So far we have only considered one-dimensional (1-D) data capture and analysis. Two-dimensional (2-D) data are very widely used to represent images from cameras, microscopes, telescopes and so forth. A 'pixel' (short for 'picture element') is the name given to the 2-D equivalent of a sample in 1-D data. A sample in a 1-D array is specified by its position in the array (sample number or time) and its value represents the amplitude of the signal at that point. A pixel is specified by its position in the image (usually by its x and y coordinates) and its value represents the brightness of the image at that point. For colour work each pixel still has a unique position but a collection of values representing the brightness of the red, blue, and green component of the image at that point.[12] Images that are essentially monochrome can have false colour applied: different colours are assigned to each grey level. One common application is for satellite images. An infrared satellite image is monochrome but the addition of false colour makes it easier to appreciate the details of the image.

Two-dimensional image analysis is beyond the scope of this book and only a few ideas will be covered here. An excellent book on the subject is *The Image Processing Handbook* by John Russ. There have been great improvements in the software available for image analysis in recent years because ordinary personal computers now have the computing power and memory needed to run them. Image-Pro Plus and Northern Eclipse are two commercial packages that can be used in a broad range of image analysis applications. Matlab has add-on 'toolboxes' for image analysis, and National Instruments has brought out a complete range of hardware and software for image capture and analysis. There is also a good image analysis program (NIH Image) that is not only well supported but is in the public domain.[13] Software packages such as Adobe Photoshop, designed for the graphic arts market, are very useful for some scientific imaging tasks.

All the signal-processing functions that are used on 1-D data can also be used on 2-D images. Because we are so used to looking at images and recognizing shapes, it can be difficult to visualize what a high-pass filter, for example, will do to an image. In a 1-D array, the high frequencies are contained in the parts of the waveform that change amplitude quickly. In a black-and-white image the amplitude of the signal is coded as brightness. The high-frequency parts of an image are therefore the parts where the brightness changes quickly. The term *spatial frequency* is used in image analysis because the frequency of an image is the rate at which the amplitude changes with distance across the image. Figure 5.59 shows an image before and after high-pass-filtering (a uniform grey level has also been added to the filtered image to avoid negative values). The areas with high spatial frequency (the facial features on the cat and the junction between the snow and the cat's fur, for example) have been preserved whereas the low-frequency information in the snow is lost.

Two-dimensional image processing makes heavy demands on computing power because of the number of calculations needed. One common way of speeding up the process is to use only integer arithmetic. This means that the filters used are often surprisingly crude when compared to those used on 1-D data, because all the coefficients are rounded to the nearest integer. For example, the high pass filter used in Figure 5.59 has the coefficients listed in Table 5.4.

(a)

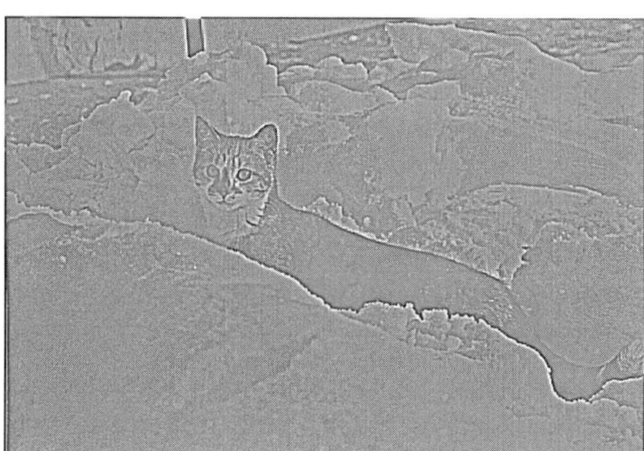

(b)

Figure 5.59 High-pass filtering in two dimensions: (a) original image; (b) image after high-pass filtering. The parts of the original image with high spatial frequencies, that is, rapid changes in brightness, are preserved.

The filter, like the image, is two-dimensional, and the image is filtered by convolving it with the filter coefficients. Filtering each pixel in the image requires twenty-five multiplications and additions, whereas the same filter applied to 1-D data would only require five multiplications and additions (the filter coefficients used for one dimension would be the five numbers in the middle row of Table 5.4).

Spectral analysis can also be performed on images, using a 2-D FFT or FHT. Spectral techniques are often used for lossy image compression: the frequency components are stored rather than the original image. The JPEG (Joint Photographers Expert Group) compression standard uses the discrete cosine transform (similar to the Fourier transform) to convert small blocks of the image into frequency components. The number of frequency components that are retained is determined by a 'quality' factor that trades compression ratio against loss of image quality. Further compression techniques are applied to the coefficients before they are stored, with the result that the image can be compressed by ratios of 10:1 up

TABLE 5.4 THE COEFFICIENTS OF THE TWO-DIMENSIONAL HIGH-PASS FILTER USED IN FIGURE 5.59

−1	−1	−1	−1	−1
−1	−1	−1	−1	−1
−1	−1	24	−1	−1
−1	−1	−1	−1	−1
−1	−1	−1	−1	−1

to 100:1. Wavelet transforms are also becoming important in image manipulation because they can be used in a manner similar to the FFT to compress images but the distortion caused by loss of image information is less obvious to the eye.

A different application of 2-D image analysis is in systems that track the position of an object. These systems are used in behavioural research to track the motion of animals, and in kinesiology to track the position of the limbs in a moving subject. The basic idea is to use a video camera to track the position of a marker on an animal or limb in motion. The first task is to make the object of interest stand out against the background, that is, produce high contrast. When small rodents are being tracked, this is achieved by making the background colour different from that of the animal. For example, tracking systems work well when black mice are photographed against a white background. In kinesiology research the contrast is achieved by sticking small reflective disks over the points of interest which are usually limb joints. Next, a video camera is used to take monochrome pictures of the object and background. The images are digitized and run into the host computer. The first and most important task in processing the data is to find the x- and y-coordinates of the animal or marker. A single video picture or 'frame' is made up from a number of horizontal lines that fill the screen. To digitize a complete frame, each line is digitized in turn. The line is sampled at a fast rate and each digitized value corresponds to the brightness of the image at that point on the line.[14] The end result of digitizing a frame is a matrix of numbers which correspond to the brightness of the individual pixels. The x-coordinate of a particular pixel is the position (left to right) of the pixel on its line, and the y-coordinate is the number of the line in the frame. To locate the coordinates of the marker, we have to search the matrix for a collection of pixels that have a value different from the background. This is where the high contrast is important. We want the pixels that represent the marker to be easy to find, and the higher the contrast the easier they will be to find. For example, suppose that black is represented by a value of 0 and 255 represents white.[15] A black mouse on a white background can be located by searching for pixels that have a value close to 0 in a background of pixels that have values close to 255. In practice, we set a cut-off value so that all pixels below this value are considered black and all pixels above it are considered white. The cut-off value is determined by looking at the whole image and looking for the value that best divides the pixels into two groups with least overlap. A typical value would be around 50, so that black and dark grey pixels are lumped together as black.

Once all the black pixels have been located we need to make sure that they represent the mouse and not just noise. The pixels are examined to see which ones are close together. Adjacent black pixels are grouped together, whereas scattered black pixels that are not adjacent are considered noise and ignored. With luck, there will be just one group of pixels which represent the mouse. Finally, the centre of the group of black pixels has to be estimated. This position is taken to be the location of the mouse. The calculations needed to locate the mouse are not complex but they have to be completed rapidly. Typically, a digitized video image is 640 pixels wide by 480 pixels high. This matrix of pixels has to be processed in less than 1/30th of a second (1/25th of a second in Europe) in order to keep up with the frame rate of a video camera.

The second stage of processing involves taking the stream of x,y-coordinates of the mouse and using them to quantify the movement of the mouse. The distance that the mouse moves between frames is easy to calculate by assuming that the mouse has moved in a straight line. If the x,y-coordinates of the mouse in the two frames are x_{old}, y_{old} and x_{new}, y_{new} then the distance that the mouse has moved is

$$\sqrt{(x_{new} - x_{old})^2 + (y_{new} - y_{old})^2}$$

The time between frames is known so the velocity of the mouse can be calculated. The position of the mouse can also be examined to see how much time the mouse spends in different parts of the image, which indicates how much the mouse explores its surroundings. The proportion of time that the mouse spends moving around indicates how active the mouse is. Activity is a good marker of normal versus abnormal behaviour in small rodents and is used in several research areas. Rats and mice are nocturnal animals and most of their activity is at night, and hence an objective, automatic activity monitoring system is desirable because the alternative is to spend the night recording animal activity by hand.

Video-tracking systems used in kinesiology are more complex because several markers have to be tracked at once and multiple cameras may be used. Merging several video images and tracking multiple markers are difficult tasks for any computer and the data may have to be processed offline.

Video tracking systems also have to be calibrated before use. The computer works on a matrix of numbers which represent digitized pixels. The distance between two pixels in real space has to be determined during a calibration step. Two markers are placed a known distance apart in the field of view of the camera. The computer calculates their separation in pixels and works out the corresponding calibration factor. The calibration has to be performed in both the x- and y-directions because the spacing between pixels on a line (x-direction) is usually not the same as the spacing between lines (y-direction). More sophisticated calibration systems may be needed if the camera lens distorts the image. Ideally, the camera will record a 'flat' image where the calibration factor is the same everywhere. In practice, the image may be distorted by the lens. This is more likely when wide-angle lenses are used. The system now has to be calibrated by placing a grid in the field of view. The lens distortion bends the straight lines into curves, producing a 'barrel' or 'pincushion' effect. The computer knows what shape the grid should be and calculates an appropriate reverse function that 'bends' the lines straight again.

Three-dimensional (3-D) signals are also fairly common in life sciences, with medical imaging being an important application. Computer-assisted X-ray tomography (CAT scanning) essentially produces a 2-D image. A 3-D image is constructed by moving the patient or the imager between images to obtain a stack of 2-D 'slices'. These are then combined in software to form a 3-D object that can be viewed and 're-sliced' from any direction. Magnetic resonance imaging (MRI) produces 3-D images either from a stack of slices or by directly imaging the entire volume. Atomic-force microscopes and tunnelling electron microscopes produce a 3-D representation of a surface with resolution in the angstrom (10^{-10} m) range.

Confocal microscopy also produces a series of slices that are recombined into a 3-D object. There are some image enhancement techniques used in confocal microscopy (deconvolution of the point spread function, for example) that require accurate convolution in three dimensions. Not surprisingly, these processes are computationally very intensive and need fast processors to complete in a reasonable time.

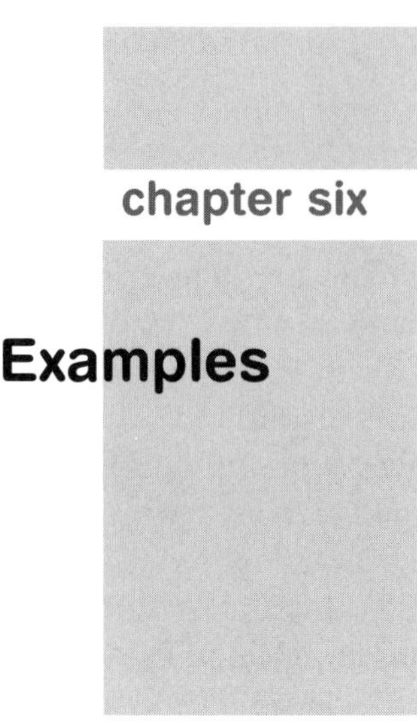

chapter six

Examples

In this chapter some examples of signal-processing problems are presented to show how the various techniques can be applied to data to extract useful information and reject artefacts. Details of the individual techniques are in Chapter 5.

ECG SIGNAL

Figure 6.1 shows part of an ECG signal recorded at a sampling rate of 1 kHz with a run-of-the-mill data acquisition system. This sampling rate should be well above the Nyquist frequency as ECG signals contain little information above about 100 Hz. The signal has been amplified 2,500 times so that a 1 mV signal appears as 2.5 V. Two problems with the signal are immediately apparent:

1. There is some high-frequency noise in addition to the ECG signal. It is very likely that this is 60 Hz interference from the mains wiring.
2. A DC offset of about −2 V is present: the ECG signal should be zero in between beats. The offset is generated by the interaction of the electrodes and the skin, forming an electrolytic half-cell. Normally it is filtered out by the ECG preamplifier but in this case a DC coupled amplifier was used.

Both the problems above can be reduced by altering the recording setup and using good quality silver–silver chloride electrodes but in many cases signals are recorded under less than ideal conditions and artefacts such as these are common. The first task is to remove the artefacts and clean up the signal.

The DC offset is removed by calculating the mean value of the signal and subtracting it from every point. It is often much easier to process a signal by breaking it up into manageable chunks, rather than trying to perform calculations on millions of samples. In this case the segment is 12.6 s long and includes twenty

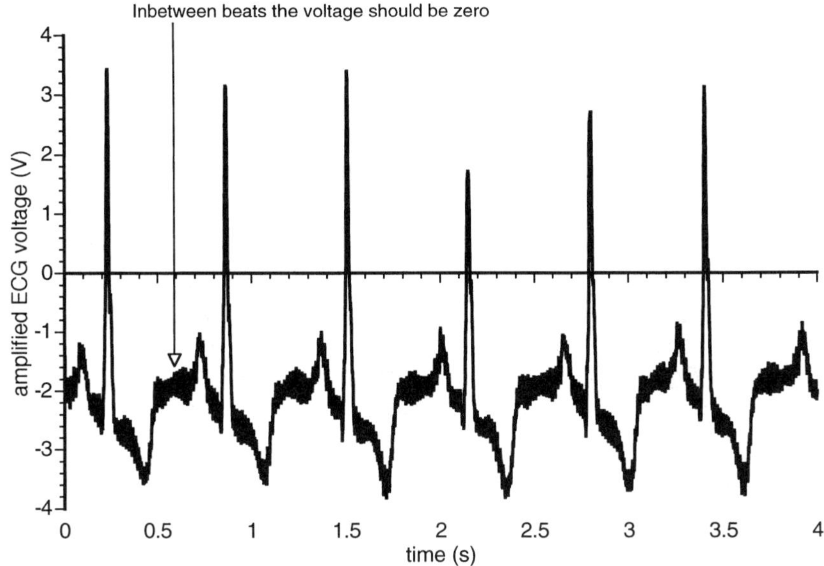

Figure 6.1 An amplified ECG signal sampled at 1 kHz. High-frequency noise and a DC offset are also present.

beats. The mean value is $-1.912\,$V. Subtracting this gives the signal in Figure 6.2 and the ECG now returns to zero in between beats.

The next step is to remove the high-frequency noise. We can try to measure the frequency of the noise directly from the recording by zooming in and measuring the time between successive peaks. But a better approach is to jump right in and calculate the spectrum of the signal. Figure 6.3 shows the magnitude of the FFT of the first 4,096 points in the segment using a Hamming window. Only data up to 100 Hz are shown (the spectrum extends up to the Nyquist frequency, 500 Hz) because there is nothing except a little noise beyond 70 Hz. This tells us that the signal is not aliased. If it were, there would be significant signal present right up to the Nyquist frequency. The spectrum shown in Figure 6.3 was calculated after the DC offset was removed from the data and the amplitudes of the first and second bins are 0.194 V and 0.111 V, respectively. For comparison, the amplitudes of the same bins without prior DC removal are 2.259 V and 0.990 V.

There is a peak at 60 Hz, as expected, but also another peak at 67 Hz. There are also smaller peaks (not shown) at multiples of 67 Hz (134 Hz, 201 Hz, etc.). The 60 Hz peak is due to mains interference. The 67 Hz peak is from a different source, the vertical refresh of the computer monitor. Computer monitors generate appreciable electrical noise and there is a strong component at the screen refresh rate: 67 Hz is a typical rate. The preamplifier was located too close to the monitor

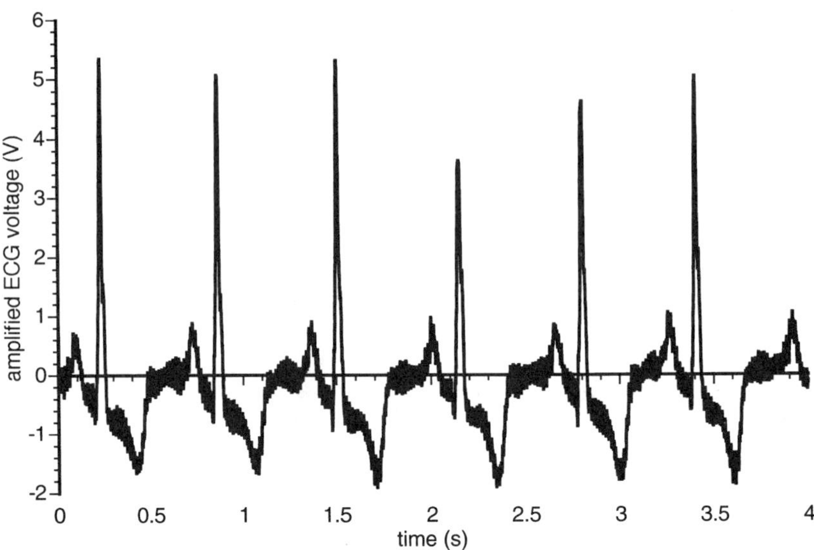

Figure 6.2 The ECG signal from Figure 6.1 with the DC offset removed.

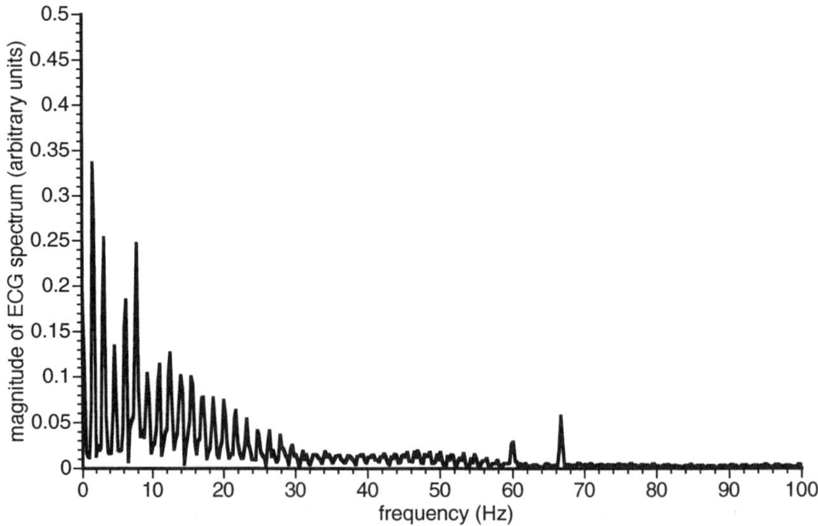

Figure 6.3 Magnitude of the spectrum of the ECG signal after offset DC removal. The peaks at 60 Hz and 67 Hz are noise picked up by the data acquisition system.

and picked up some of the interference. Another clue is the presence of harmonics of the 67 Hz signal (i.e. components at multiples of 67 Hz), which are caused by the spiky nature of the refresh signal. The 60 Hz mains interference is a fairly pure sine wave – there is no harmonic at 120 Hz.

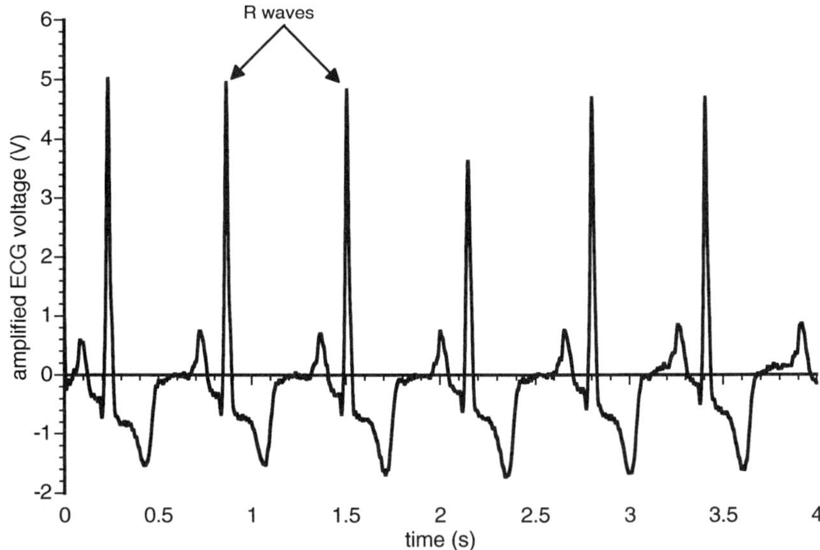

Figure 6.4 The ECG signal from Figure 6.2 after low-pass filtering with a corner frequency of 45 Hz.

The easiest way to get rid of the interference is to filter the data. A notch filter, which removes signal components in a narrow frequency band, could be used with frequency limits of about 55–75 Hz. On the other hand, there is some ECG information above 60 Hz but not much and for many applications it is not important. A simple low-pass filter will therefore suffice in many cases. To give some leeway the corner frequency of the low-pass filter should be a bit below 60 Hz. Figure 6.4 shows the signal after applying a low-pass FIR filter with a corner frequency of 45 Hz, and Figure 6.5 the corresponding amplitude spectrum. We now have a nice clean ECG that is suitable for further processing. At this point the ECG signal can be compressed considerably by resampling with a Nyquist frequency greater than 45 Hz, since everything above this has been filtered out. A sensible choice would be a sampling rate of 150 Hz giving a Nyquist frequency of 75 Hz.

A very common aim in ECG processing is to identify the R wave (Figure 6.4) so that the heart rate can be calculated. The literature is full of R wave identification algorithms (e.g. Pan and Tompkins 1985), many of which are quite complex and unique but are not available in standard signal analysis programs. Here we will use readily available processing algorithms. The most straightforward way of locating the R waves is to use a peak detector. We can set fairly wide hysteresis bands and count the peaks that emerge out of the top of the band. Figure 6.6 shows

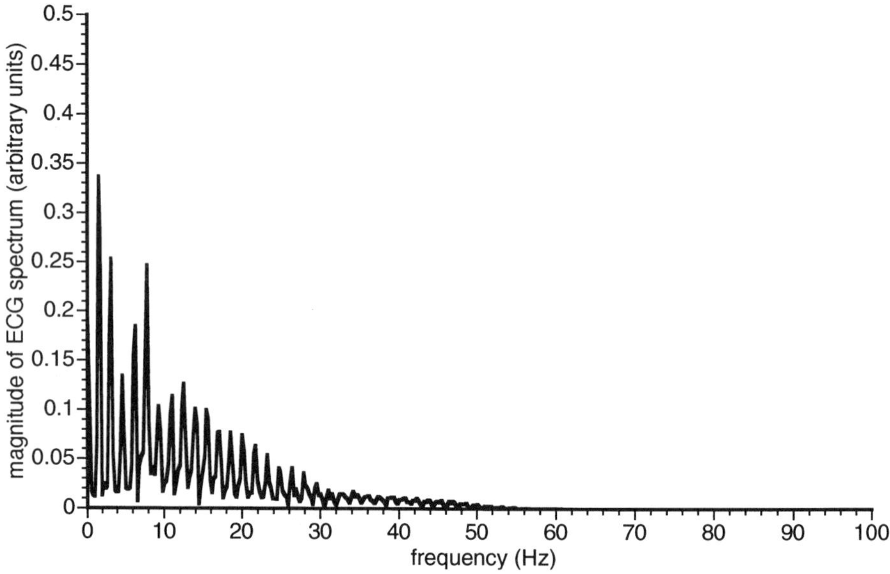

Figure 6.5 Magnitude of the spectrum of the filtered ECG signal in Figure 6.4. Signal components above 55 Hz have been reduced to insignificant levels.

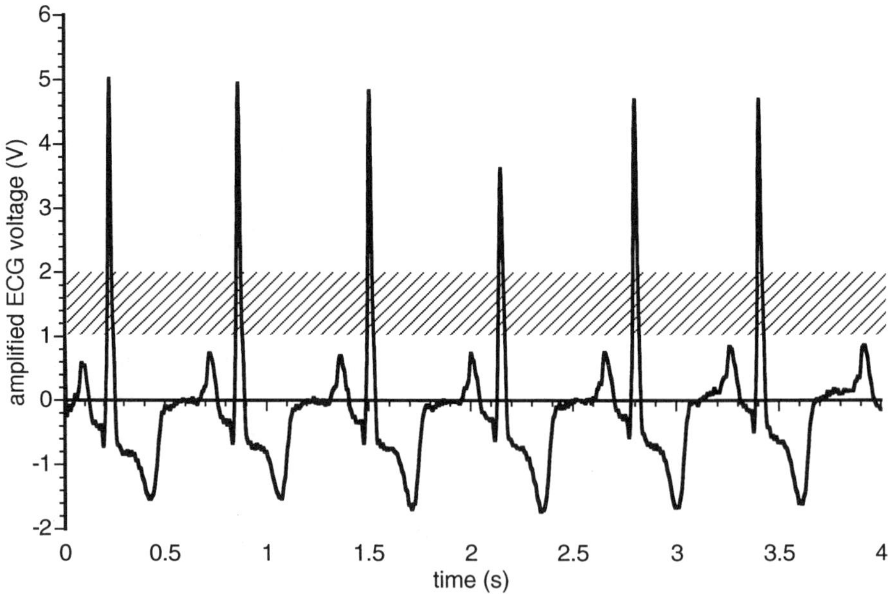

Figure 6.6 The filtered ECG signal from Figure 6.4 with a superimposed hysteresis band (diagonal hatching) from 1.0 V to 2.0 V.

the ECG signal with a superimposed hysteresis band of 1.0–2.0 V. The algorithm moves along the signal looking for a point where the signal emerges from the top of the hysteresis band. From here to the time that it reenters the band is regarded as a single peak, with a position corresponding to the highest point in the section. Using this algorithm all the R waves in this recording were detected with no errors. There are, however, several problems with this simple method of R-wave detection. Firstly, the hysteresis bands are sensitive to the signal amplitude. If the R-wave height falls, there will come a point where no R waves are detected at all. Secondly, the waveform has to have a positive R wave. If the electrodes are reversed (giving an inverted signal), no R waves will be detected. Thus the algorithm is sensitive to the size of the patient and the electrode position, which makes it very inflexible.

One method of R-wave detection that is widely used is to look for the rapid change in voltage at the start and end of the R wave. Here we are interested in the rate of change of voltage, not the absolute level. If the ECG signal is differentiated the result is proportional to the rate of change of voltage and the rapidly changing sections will be accentuated. As we are differentiating, it is advisable to first low-pass filter the data and the 45 Hz filter that has already been applied will suffice. The differentiated signal is shown in Figure 6.7. The peaks, corresponding to the

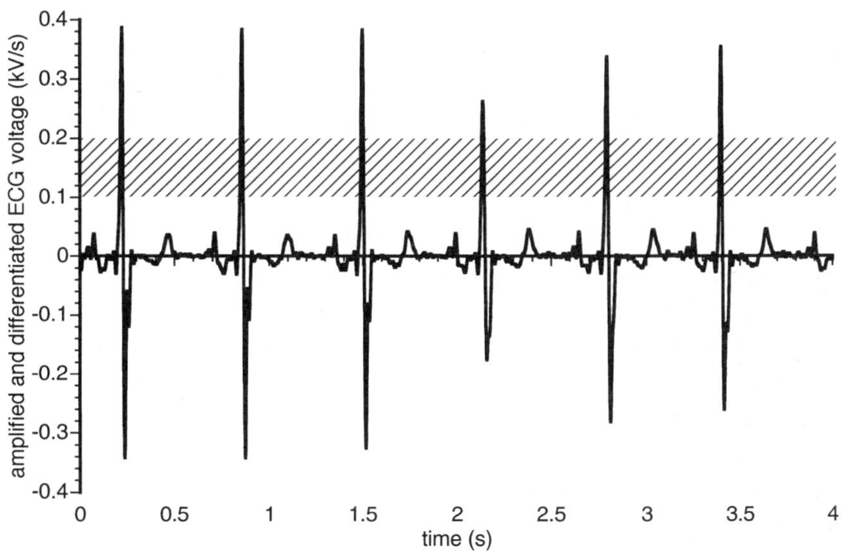

Figure 6.7 Differentiated ECG signal with superimposed hysteresis band (diagonal hatching) from $0.1\,\mathrm{kV\,s^{-1}}$ to $0.2\,\mathrm{kV\,s^{-1}}$. The single positive R-wave peaks have been converted into bipolar peaks.

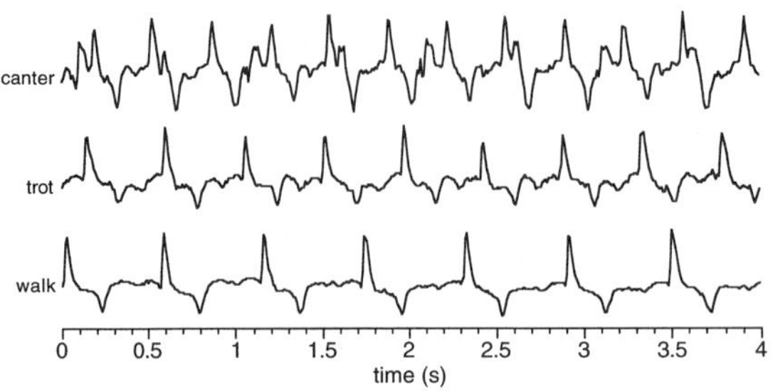

Figure 6.8 An ECG signal recorded from a horse at three different paces: walk, trot, and canter. As the speed increases so does the amount of interference from skin movement and muscle EMG.

rapidly rising and falling parts of the R waves, can now be located using the same peak-detection algorithm as described above. The difference is that, by first differentiating the signal, the R-wave detection process is made independent of the signal polarity. The R wave is quite symmetrical and thus it rises and falls at similar rates, giving equal positive and negative peaks in the differentiated signal. The R wave detector is also less dependent on the ECG amplitude because the differentiated R wave is now larger relative to the other components of the ECG and the peak-detector limits can be set quite far apart. Differentiation followed by peak detection is commonly used in commercial ECG monitors for locating the R wave.

The ECG signal in Figure 6.1 is quite 'clean', that is, there is not much extraneous noise. In many cases we have to deal with far higher levels of noise. A common example is recording an ECG from an exercising subject – the ECG is contaminated with high levels of EMG from the muscles which can only be partly alleviated by siting the electrodes away from muscle masses. Figure 6.8 shows such an ECG from an exercising horse. At the walk the ECG complexes are clear and there is little noise. As the horse speeds up into a trot and then a canter, the interference increases so that at the canter some of the noise spikes are almost as large as the R waves. The individual complexes can still be recognized by eye but both the algorithms described above fail to reliably detect the R waves. Given such a signal our main aim is just to detect the presence or absence of an R wave – the waveform is too distorted to make any meaningful interpretation of its shape. Detecting the presence of a short waveform in a longer recording is an ideal task for a matched filter. It can be shown that a matched filter is the optimum detector for a pulse – there is no point in performing any further filtering of the data before the matched filter because it will not increase the signal-to-noise ratio. A matched filter requires a 'template' to work from. Figure 6.9 shows a single waveform taken from the filtered signal in Figure 6.4. This waveform is the template. The waveform is then reversed in time (Figure 6.10), which simply involves

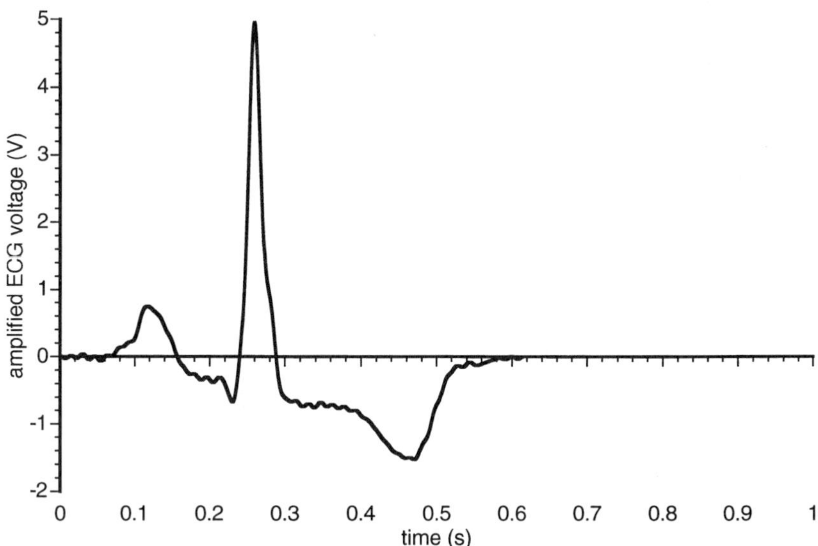

Figure 6.9 A single ECG complex from Figure 6.4, to be used as a 'template' for a matched filter.

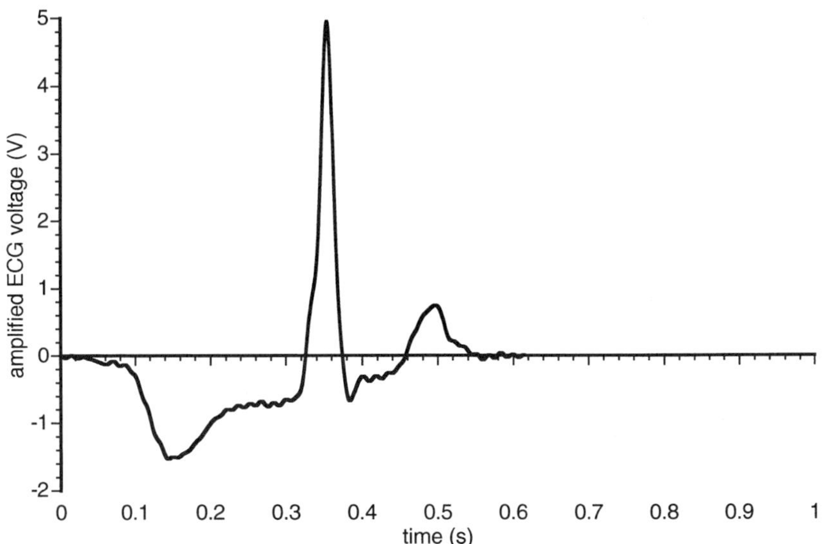

Figure 6.10 The ECG complex from Figure 6.9 reversed in time.

arranging the samples in reverse order. Next, this reversed pulse is cross-correlated with the original signal. The mechanics of cross-correlation are explained in Chapter 5, but there is a very simple way of implementing it. Most programs will allow you to load custom FIR filter coefficients. Use the time-

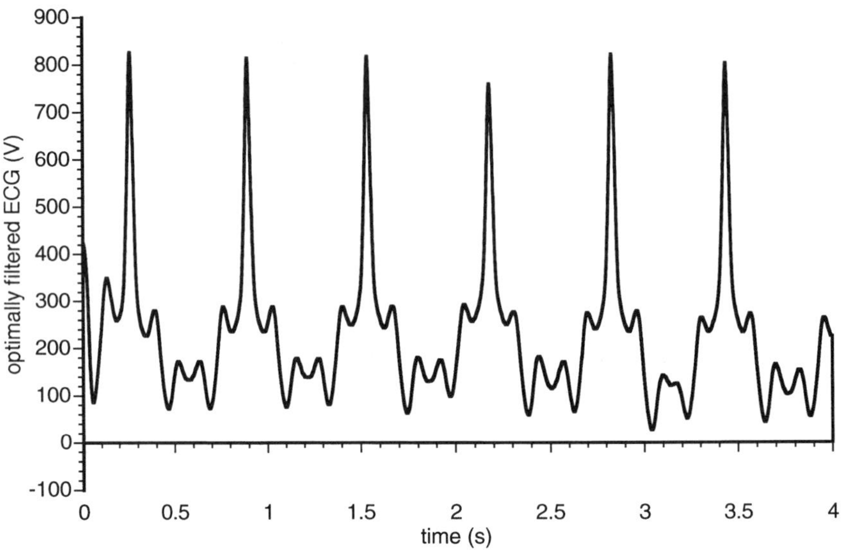

Figure 6.11 Matched filtration: the result of cross-correlating the reversed complex with the original ECG signal. The shape of the ECG waveform is lost and now each peak corresponds to the position of an ECG complex in the original signal.

reversed waveform as a set of FIR filter coefficients and filter the waveform. Figure 6.11 shows the result of doing this to the original waveform in Figure 6.1. Note that the matched filter does not preserve the amplitude or the shape of the ECG waveform. Instead, the amplitude of the filtered signal is proportional to how well the signal matches the template so that there is a peak each time the template is aligned with a QRS complex. Again, a simple peak detector is then used to locate each R wave. The power of the matched filter can be seen by taking the original ECG waveform and adding a large amount of random noise to it (Figure 6.12). The original ECG complexes are now barely recognizable but when the matched filter is applied (Figure 6.13) the R waves can still be detected reliably. The drawback to the matched filter is that the template should be as similar to the signal to be detected as possible and thus the detector will not work well if the signal is inverted. However, having a fixed lead configuration is a small price to pay for the ability to detect the ECG complexes in the presence of such large amounts of noise.

CARDIAC OUTPUT BY INDICATOR DILUTION

The commonest way of measuring cardiac output (the amount of blood that the heart pumps in one minute) in clinical medicine is by indicator dilution. The principle is straightforward. Imagine a pipe with water flowing through it

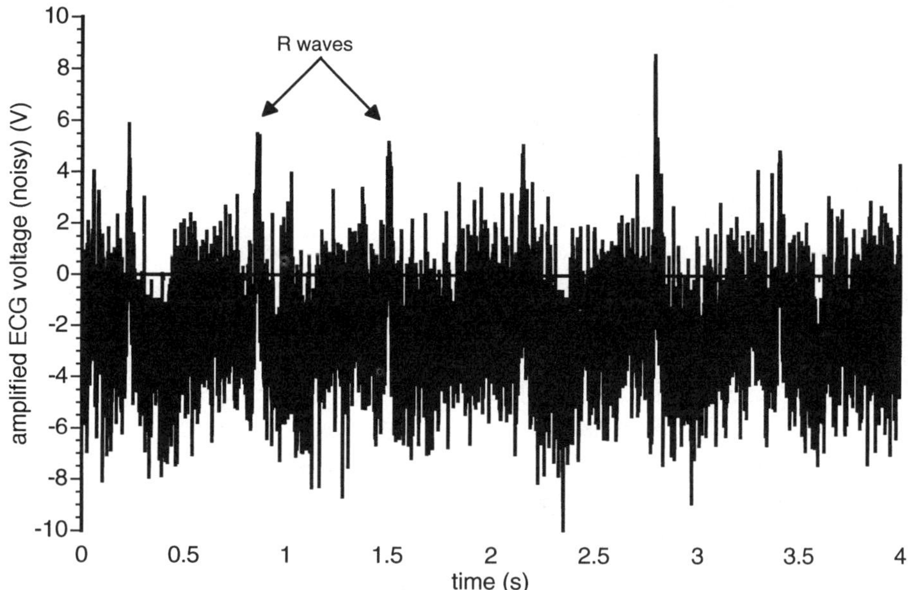

Figure 6.12 The ECG signal with a large amount of random noise added. The ECG complexes are barely recognizable.

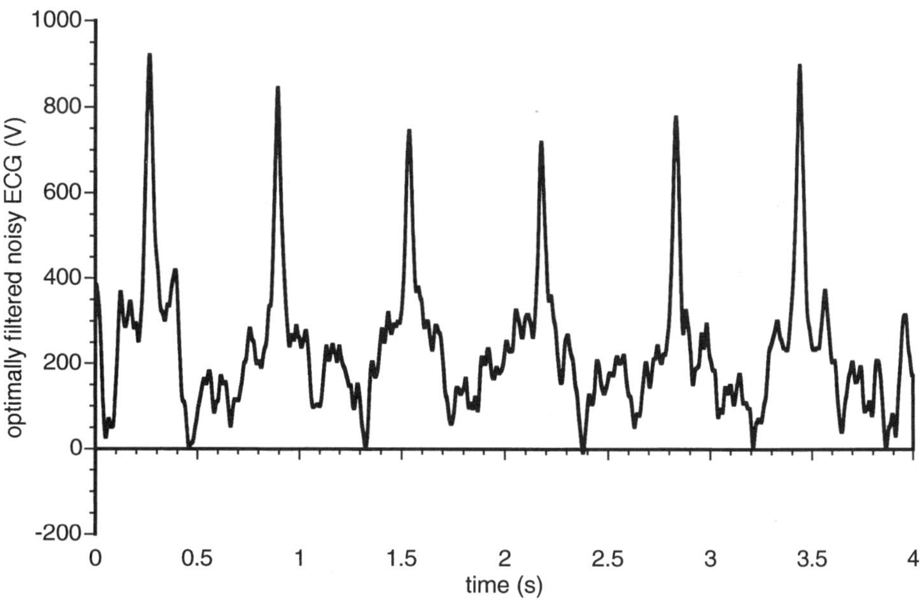

Figure 6.13 The result of matched filtration of the noisy ECG signal. Each large peak clearly marks the position of an ECG complex.

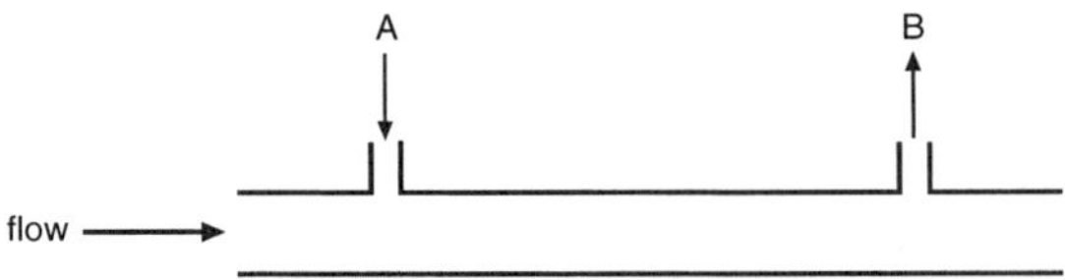

Figure 6.14 The principle behind flow measurement by indicator dilution. An indicator is injected at point A, mixes with the main flow and its concentration is measured at point B. Knowing the injection rate the flow can be calculated.

(Figure 6.14). We want to measure the flow rate of the water. At point A, an indicator (a substance such as a dye whose concentration in the water can be measured readily) is injected into the pipe at a known rate. At a second point B downstream, the indicator has mixed thoroughly with the water, and the concentration is measured. Knowing this and the rate of indicator injection the flow rate of the water can be calculated. The calculation depends upon the assumption that indicator is not lost from or added to the pipe so that the amount of indicator injected at point A is the same as that passing point B:

$$\text{rate of injection of indicator} = a\,(\text{mg s}^{-1})$$
$$\text{concentration of indicator at point B} = b\,(\text{mg ml}^{-1})$$
$$\text{flow rate} = Q\,(\text{ml s}^{-1})$$

The rate that indicator passes point B $= b \times Q\,(\text{mg s}^{-1})$. But this is the same as the rate of injection, a. Therefore,

$$a = b \times Q, \text{ or } Q = \frac{a}{b}\ (\text{ml s}^{-1})$$

In the practical application of this technique to measure cardiac output, the indicator is not infused continuously but a single injection is given, and instead of the final concentration the area under the concentration–time curve is used to calculate cardiac output. A nontoxic green dye (indocyanine green) can be used as the indicator and the concentration in the blood measured optically. Figure 6.15 shows the concentration of dye in the arterial blood (sampled at 10 Hz) after injecting 10 mg of dye intravenously into a 400 kg horse. The cardiac output is calculated from the equation

$$\text{cardiac output (l/min)} = \frac{\text{amount of dye (mg)} \times 60}{\text{area under concentration–time curve (mg l}^{-1}\,\text{s)}}$$

Before the calculation can be made, the data have to be tidied up. The first task is simply to establish a baseline. The calculation of cardiac output assumes that

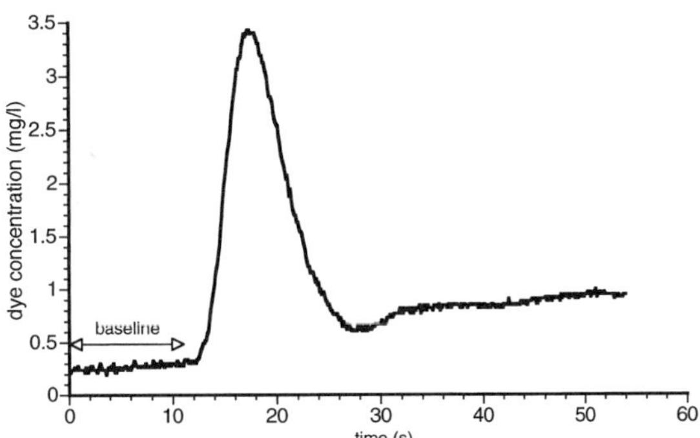

Figure 6.15 The concentration of green indicator dye in the arterial blood of a 400 kg horse after an intravenous injection of 10 mg dye. Linear regression is now used to fit a line to the baseline portion for de-trending.

there is no dye present before the injection, yet the graph shows that there is about $0.25\,\mathrm{mg\,l^{-1}}$ dye in the blood at the start of the recording. This background dye concentration is left over from previous measurements and must be removed. The simplest solution is to select a part of the trace just before the concentration starts to rise ('baseline' section, Figure 6.15), calculate the average value ($0.265\,\mathrm{mg\,l^{-1}}$) and subtract it from the whole curve. This is fine, but the baseline section is not really horizontal. It slopes upwards, indicating that there is some drift present as well. This can be removed by fitting a straight line to the baseline section (linear regression) and subtracting that from the whole curve. Linear regression gives the equation of the straight line that best fits the data:

$$y = 0.0062x + 0.233$$

where y is concentration and x is time. To subtract this from the curve the formula is used to calculate a y value for each sample and subtract that from the measured concentration. The result is shown in Figure 6.16. Note that this technique would have also worked if the baseline was in fact horizontal. Subtracting a best-fit straight line from a set of data in this fashion is called *de-trending*.

The next task is to correct for recirculation of the dye. At point A (Figure 6.16) the concentration rises again slightly as the dye comes round a second time, and this recirculation effect has to be removed in order to obtain a true measurement of cardiac output. The standard way of doing this is to extrapolate the falling 'wash-out' section of the curve (Figure 6.16) back down to the baseline. The wash-out section is assumed to have an exponential shape, which has some justification because many wash-out processes follow an exponential curve quite accurately.[1] Traditionally the extrapolation was performed by plotting the data on logarithmic paper so that the exponential curve became a straight line, but it is easier just to fit an exponential curve to the data. Two points on the falling section of the curve,

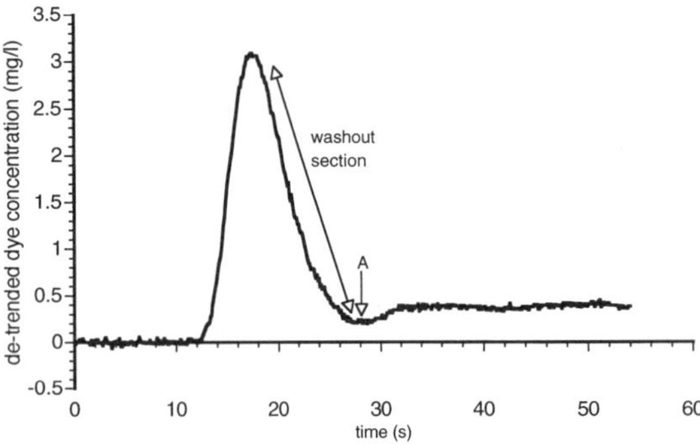

Figure 6.16 The dye curve from Figure 6.15 after de-trending. At point A re-circulation of the dye starts. The wash-out section starts when the dye concentration begins to fall and extends to the start of recirculation.

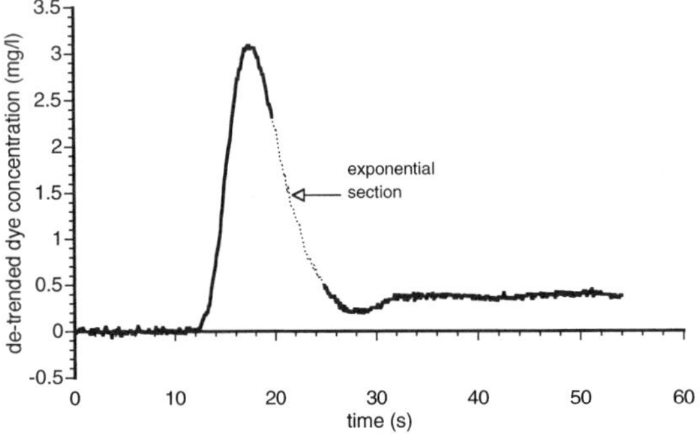

Figure 6.17 The central part of the wash-out section is selected for exponential curve fitting.

about one-third and two-thirds of the way from the peak to the baseline, are selected (Figure 6.17). An exponential curve is fitted to the data between these points. In this particular case the curve fit yields the equation

$$\text{concentration} = 1021.7\text{e}^{-0.3065t}$$

with an r^2 value of 0.994, indicating a good fit. Now the equation is used to extrapolate the curve from the end of the exponential section to the baseline. This is an exception to the rule that curves should not be extrapolated beyond the data set, and is only valid because we are reasonably sure that the data follow an exponential curve. Using the equation of the curve, new concentration values can be calculated at each time interval and used to replace the original ones (Figure 6.18).

The area under the curve can now be calculated using the trapezium or Simpson's rule. Using the trapezium rule is simplest and we can also drop the factor of 1/2 for the first and last samples with good justification because the curve starts and ends close to zero. The sampling rate is 10 Hz, so the area is 0.1 × the

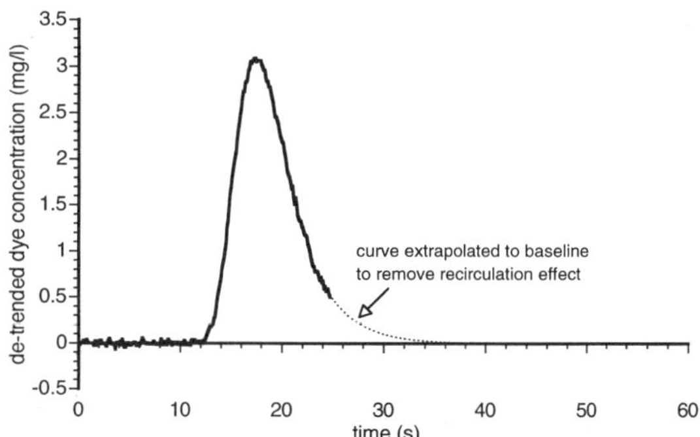

Figure 6.18 The curve is extended back to the baseline using values calculated from the curve fit to remove the effects of recirculation.

sum of all the samples. In this example the area is $22.23\,\mathrm{mg\,l^{-1}\,s}$. The cardiac output can now be calculated using the equation above, giving a value of $27.0\,\mathrm{l\,min^{-1}}$. This is a typical value for a 400 kg horse at rest.

Although the calculations involved in this example are straightforward, writing a robust program to automate the process is not a simple task. Several points on the curve have to be identified reliably and automatically: the baseline region, the peak of the curve and one-third and two-thirds of the way from the peak to the baseline. Another problem that will occur is that the extrapolated curve may not always reach the baseline. One solution is to extrapolate the data for a reasonable time (e.g. 20 s) and see how close it has come to the baseline. If it is close then the small error in the area can be ignored. If it is a long way off, the curve should be rejected. These types of problem occur all the time in signal analysis, and the art of writing good programs is to anticipate the problems and take sensible steps to deal with them. In many cases limits are set within which 'good' signals fall and signals outside the limits are rejected. There are usually strict rules about how to set the limits and it is up to the programmer to make sensible choices. For example, how close should the extrapolated curve come to the baseline for the curve to be acceptable? One approach would be to calculate 2–3% of the peak height of the curve and use this as the acceptable limit. This system has the advantage of being autoscaling so it will work for curves of different heights.

SPECTRAL EDGE FROM EEG DATA

An electroencephalogram (EEG) is a noninvasive recording of the brain's electrical activity from electrodes applied to the skin over the skull. EEGs are useful for investigating CNS disorders such as seizures, where the focus of the seizure can be determined from multilead EEG recordings. The general activity of the brain can also be determined from the EEG by spectral analysis. The power spectrum of the

EEG is calculated and divided into 'waves', which are in reality frequency bands rather than discrete waves. The relative powers of the bands reflect the level of brain activity.

There has been considerable research into finding simple indices of the depth of anaesthesia in anaesthetized humans and animals. It can be remarkably difficult to tell how deeply a person or animal is anaesthetized, particularly if muscle relaxants have been used. The EEG of anaesthetized humans has been analysed in many ways and the *spectral edge* has been found to correlate well with anaesthetic depth. The spectral edge is the frequency that divides the total EEG power into two defined fractions. For example, the 90% spectral edge divides the total EEG power into two segments containing 90% and 10% of the total power. The spectrum from DC to the 90% spectral edge contains 90% of the total power, the rest of the spectrum contains the remaining 10%.

Figure 6.19 shows a 2 s segment of raw EEG signal recorded at 1 kHz from an anaesthetized donkey. We wish to calculate the 90% spectral edge from this recording. As this recording shows, EEG signals are fairly random in nature and difficult to analyse in the time domain, which is why spectral techniques are useful. The first task is to calculate the spectrum of the signal. It would be nice to use the FFT because it will be much faster than the DFT on this large (2,000 sample) recording, but there is a problem. The FFT is most efficient when the number of samples is a power of 2, and the nearest powers of 2 are 1,024 and

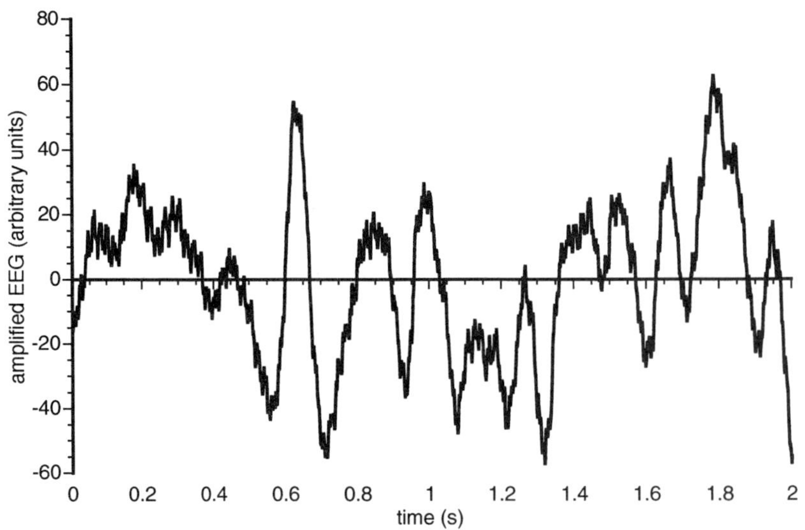

Figure 6.19 A segment of raw EEG signal recorded from an anaesthetized donkey.

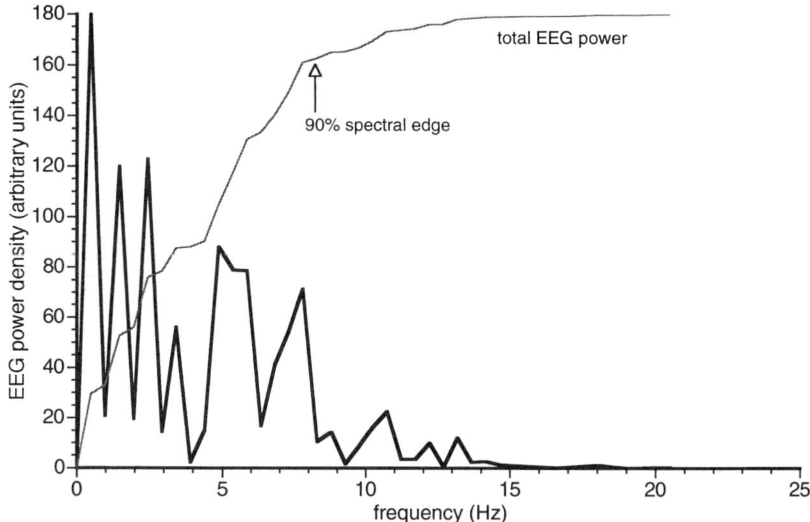

Figure 6.20 The power density spectrum of the EEG signal from Figure 6.19 (bold line) and its integral (pale line). The position of the 90% spectral edge is at 8.2 Hz.

2,048. We can throw away half of the data and calculate the FFT on 1,024 points, but this will reduce our frequency resolution to 1 Hz per bin as well as losing data. On the other hand, 2,000 is very close to 2,048 and if we 'pad' the segment with 24 zeros at each end we can calculate the spectrum on 2,048 points with little distortion of the spectrum. We are going to window the data anyway and this greatly reduces the contribution of the samples at each end of the segment.

Figure 6.20 shows the EEG power density spectrum from 0 to 20 Hz calculated by padding the original segment with 24 zeros at each end, windowing with a Hamming window, taking the FFT and squaring the magnitude of the result. The spectrum continues up to the Nyquist frequency (500 Hz) but the rest is not shown because it contained no signal apart from some power line interference at 50 Hz and a smaller contribution at the third harmonic, 150 Hz (the recording was made in the UK).[2] One consequence of zero-padding the original recording is that the bin width is 0.488 Hz rather than 0.5 Hz because the padded recording is now 2.048 s long.

Also shown in Figure 6.20 is the cumulative total power (not to scale), calculated by integrating the power density curve using the trapezium rule and ignoring the factor of 1/2 at each end. It is simply the running sum of the spectral density graph multiplied by the bin width and a scaling factor to keep it on the graph. The power at 20 Hz is taken as the total EEG power because there is negligible signal beyond this. The 90% spectral edge is the frequency below which 90% of the total power is contained, 8.2 Hz in this case.

We were able to calculate the 90% spectral edge from this recording, but many calculations were redundant because the EEG only contained significant power to 20 Hz and yet the Nyquist frequency was 500 Hz. Now that the EEG spectrum is known we can optimize the signal processing to make it faster by reducing the number of samples per recording. An anti-aliasing filter with a corner frequency of 40 Hz will not affect the EEG signal and gives a margin in case the EEG spectrum changes in subsequent experiments. The sampling rate is now set so that a high order filter is not needed. A 200 Hz sampling rate gives us a Nyquist frequency of 100 Hz, which is 2.5 times, or 1.3 octaves, higher than the filter's 40 Hz corner frequency. A fourth-order filter has a roll-off rate of 24 dB per octave so the attenuation 1.3 octaves away from the corner frequency is 31 dB. As there is very little signal at the Nyquist frequency this is adequate. There is no significant signal above 50 Hz until we reach 150 Hz when the attenuation is a whopping 105 dB. Thus a sampling rate of 200 Hz and a fourth-order 40 Hz anti-aliasing filter will be adequate and will reduce the amount of data to be stored and analysed five-fold.

The frequency resolution in this example was 0.488 Hz. The only way of reducing this is to record for a longer period of time. A 10 s recording will give a frequency resolution of 0.1 Hz, which may be more desirable. The problem now is that the system cannot respond to rapid changes in spectral edge since a new result is only available once every 10 s. A partial solution to this problem is to perform the calculations on 10 s blocks of data but to overlap successive blocks. For example, by using 5 s of new data and the last 5 s of the previous block we can update the spectral edge every 5 s with a resolution of 0.1 Hz. The overlap in this case is 50%. Using overlapped data has the effect of smoothing the successive spectral estimates, which is often desirable. The greater the percentage overlap the more the spectral estimate is smoothed and an overlap of around 50% is common.

NONLINEAR CURVE FITTING TO A DOSE–RESPONSE CURVE

Figure 6.21 shows the change in respiratory resistance of an anaesthetized animal in response to increasing doses of histamine. We need to estimate the dose of histamine that produces a 100% increase in resistance which is usually called the PC_{100} (provocative concentration that causes a 100% increase in resistance). Provocative concentrations are widely used as an objective measure of airway reactivity, one of the components of asthma. By looking at the graph the PC_{100} is about 6 µg kg^{-1}. The simplest way to estimate the PC_{100} is to use linear inter-

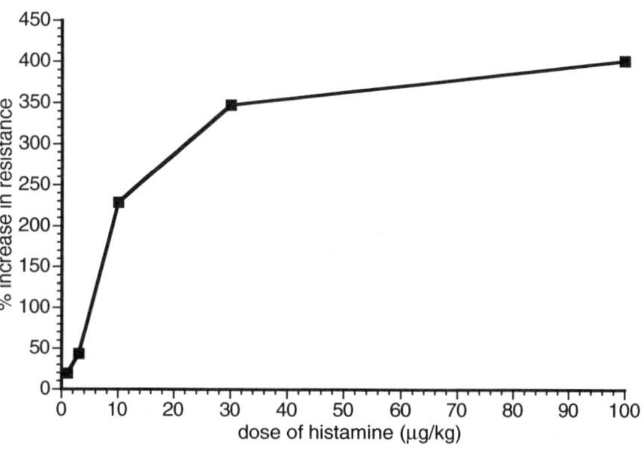

Figure 6.21 The increase in respiratory resistance of an anaesthetized animal in response to increasing doses of histamine (a dose–response curve).

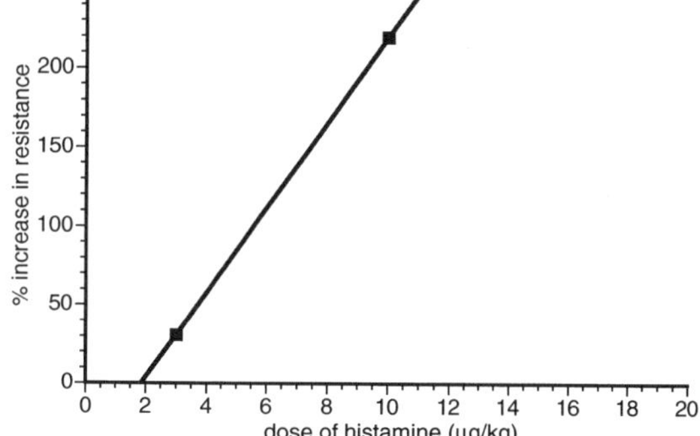

Figure 6.22 Calculation of the PC_{100} dose of histamine by linear interpolation between the two data points spanning the PC_{100} concentration.

polation, that is, draw a straight line between the two data points on either side of the PC_{100}. This is easily done by applying linear regression to the two data points (Figure 6.22). The coefficients from the regression give us the equation of the line: $y = 26.95x - 50.09$. The goodness-of-fit indicator, the r^2 value, is not very useful in this case because it is always 1; there is only one way to fit a straight line between two points. By rearranging the equation of the line to the form $x = (y + 50.09)/26.95$ the PC_{100} value is calculated to be $5.57\,\mu g\,kg^{-1}$. There are two main problems with this approach. Firstly, the dose–response curve is not linear and so an error is introduced. Secondly, the PC_{100} is determined by just two points on the curve and thus will be heavily influenced by any noise or errors associated with these two points.

Fitting a curve to the whole data set avoids both of these problems. We hope that the fitted curve better approximates the true dose–response curve than a straight line, and because this curve is calculated from all the data points the noise present in the measurements will be averaged over the whole data set.

The problem now is to select a suitable curve and fit it to the data set. Dose–response curves are usually sigmoidal in shape and there are several pharmacological models to describe them. One fairly general model is the sigmoid I_{max} equation:

$$\text{response} = \frac{R_{max}C^n}{C_{50}^n + C^n}$$

where R_{max} is the maximal response, C_{50} the concentration needed for 50% maximal response, and n is the Hill coefficient, which describes the shape of the sigmoid curve. As n becomes larger the shape of the curve changes from gentle to abrupt.

This equation is nonlinear and not in a form that is linearizable, so a nonlinear curve-fitting routine is needed. There are several commercial programs that have nonlinear curve-fitting functions and most are based on the Levenburg–Marquardt algorithm. In this case the routine in DeltaGraph was used and Figure 6.23 shows the equation fitted to the data points. The coefficients are:

$$R_{max} = 401.2$$
$$C_{50} = 8.88$$
$$n = 1.73$$

We need to rearrange the original equation in order to calculate PC_{100}. With some manipulation the equation becomes

$$\text{PC}_X = \exp\left[\frac{1}{n}\log_e\left(\frac{xC_n^{50}}{R_{max} - x}\right)\right]$$

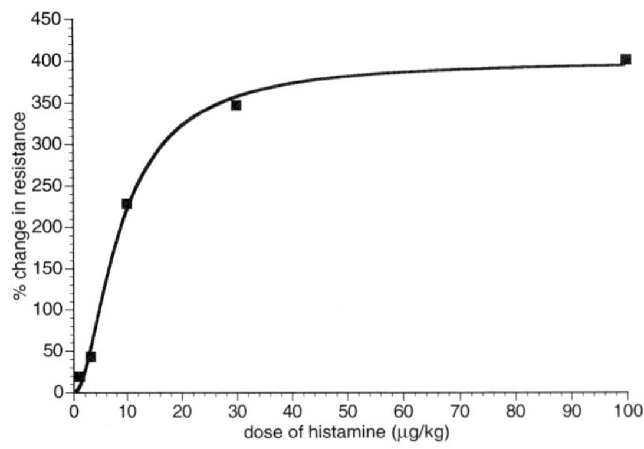

Figure 6.23 The data from Figure 6.21 with a sigmoidal function fitted.

where PC_x is the concentration of histamine required to get an $x\%$ response. Using the coefficients obtained from the curve fit, and specifying a value of 100% for x, the PC_{100} is calculated to be $4.69\,\mu g\,kg^{-1}$. In a similar manner the dose to produce any desired response can be calculated but it is unwise to extrapolate outside the experimentally determined range of responses.

The curve-fitting algorithm needs a set of initial values for R_{max}, C_{50}, and n. In this case the values used were chosen fairly arbitrarily as

$$R_{max} = 500$$
$$C_{50} = 5$$
$$n = 1$$

The algorithm found a good fit from these initial values, and in fact the same initial values have been used to analyse many data sets of the same type with no convergence problems. There is no guarantee with nonlinear curve-fitting algorithms that the final curve fit is the best one but from Figure 6.23 it is clear that it cannot be much better. It is always best to check the result of a nonlinear curve fit by eye to make sure that the fitted curve is sensible and the algorithm has not become unstable.

HEART RATE VARIABILITY

In the last few years there has been considerable interest in the detailed analysis of heart rate, one of the reasons being that it is noninvasive and may give an early indication of cardiac disease. A common requirement is to measure the heart rate and calculate the relative amount of variability in the heart rate in different frequency bands. The frequency bands are typically described as low-frequency (less than $0.05\,Hz$), mid-frequency (0.05–$0.25\,Hz$) and high-frequency ($0.25\,Hz$ up to the Nyquist limit). Heart rate variation in the low-frequency band is due to blood pressure control by the renin–angiotensin system and thermoregulation, the baro-reflex causes heart rate fluctuations in the mid-frequency band and the high-frequency fluctuations are due to respiration. Besides the heart rate variation due to these physiological mechanisms there may be nonlinear 'chaotic' heart rate variation and a fair amount of noise.

Figure 6.24 shows the time in milliseconds between successive heartbeats (the R–R interval or RRi) recorded from a volunteer resting quietly. A normal value for the resting heart rate is about $72\,beats/min$, giving a RRi of $60/72 \times 1000 = 833\,ms$. There are two ways commonly used to obtain such data. In both cases the first task is to amplify the ECG signal as cleanly as possible using an ECG monitor designed

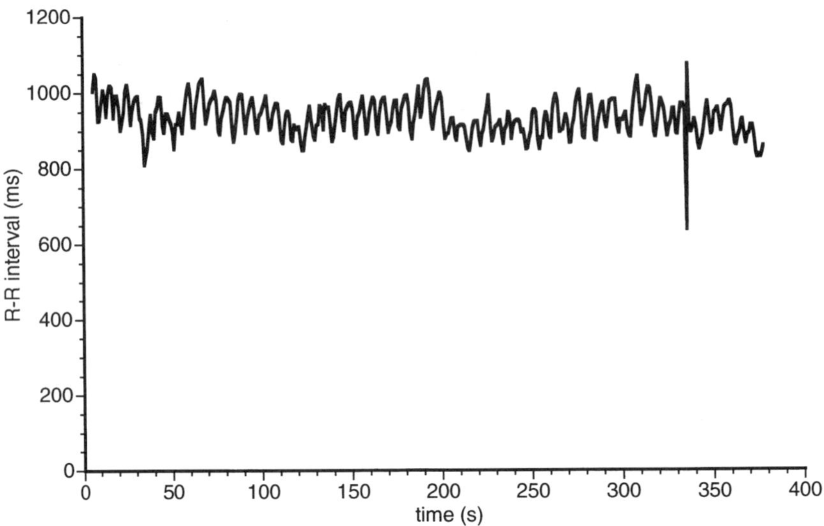

Figure 6.24 The R–R interval from a person at rest. Figure shows 400 successive beats. At 335 s there is a glitch caused by an error in the R-wave detection process which generated the data.

for the purpose, with proper patient isolation circuitry. Then the peaks of the R waves have to be found. This can be done in hardware using peak-detection circuitry, or in software using the type of algorithms described earlier. In this particular case the latter approach was taken: the amplified ECG signal was sampled at 1,000 Hz and processed later with a peak-detection algorithm. The number of samples between R-wave peaks is the R–R interval in ms.

The data in Figure 6.24 are not quite the same as conventional sampled data. The conventional approach would be to have a continuous signal that is sampled at regular intervals in time. This would mean that we would have some device that continuously measured heart rate and we would regularly sample the output of this device. Instead, the time between consecutive heartbeats is measured and so we have a recording of RRi that is sampled once per heartbeat. The heart rate is not constant, however, and thus the signal is not sampled regularly in time. Another problem is whether the sampling rate is high enough to avoid aliasing. The heart rate is fixed between beats and so it is difficult to see how sampling it more rapidly would yield any more useful information. On the other hand, the heart rate sometimes changes appreciably between successive beats and when the spectrum is calculated there is usually considerable signal energy up to the Nyquist limit, implying that aliasing may have occurred. In practice, the signal is treated as if it were sampled at regular intervals and theoretical calculations suggest that the error introduced is not that large so long as the variation in heart rate is not that large (TenVoorde, Faes, and Rompelman 1994). The question of aliasing is usually ignored as it is difficult to imagine how the heart rate could be calculated more than once per heartbeat.

The first task is to clean up the data. There are usually a few beats that have been missed by the R-wave detection algorithm because of noise from patient movement. If a beat is missed, the apparent RRi will be doubled. The algorithm may also count extra beats. Both of these artefacts are relatively easy to clean up by comparing each RRi to the mean for the whole record. Missed beats will be about double the average RRi and extra beats much smaller, and simple limits will pick up the erroneous data points. The data points can either be deleted or re-created by replacing them by the average of the RRi on either side of them. In Figure 6.24 there is a glitch at 335 s. Inspecting the data reveals that the first RRi is short (638 ms) and the second one long (1,076 ms). The sum of these two is 1,714 ms, just about right for two RR intervals so it is likely that in this case the RRi detector triggered early on some noise. The two erroneous RR intervals are replaced by their average (857 ms) to correct the error (Figure 6.25). The large DC component is then removed by subtracting the mean of the whole data series from each point to give a signal that varies about zero. Next, the power spectrum is calculated. It could just be calculated on the entire record but the resulting spectrum will be quite variable. Noise-like signals theoretically have a flat power spectrum (the spectrum is a horizontal line, that is, constant power density at all frequencies).[3]

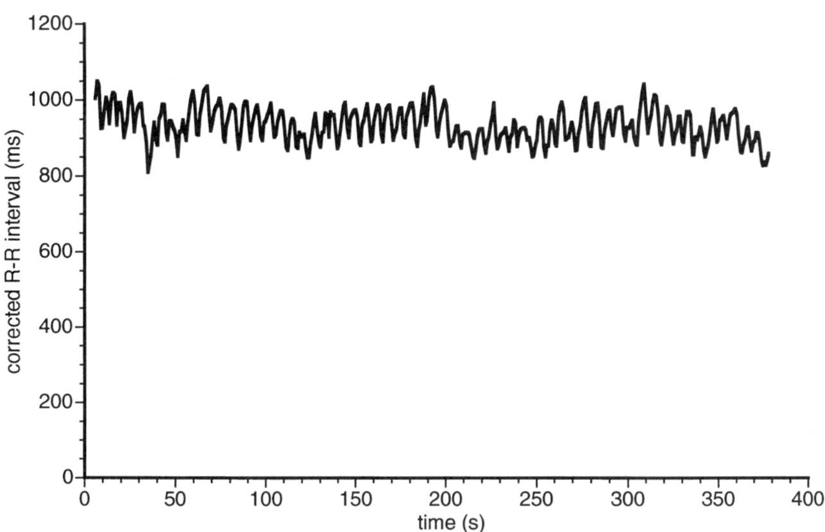

Figure 6.25 The data from Figure 6.24 with the glitch corrected.

But if you make several individual recordings of a noise-like signal at various time points and compare their spectra they will be different from each other and none of them will be close to a flat line. This is because noise-like signals are essentially random in nature and thus have to be described in statistical terms. For any single recording of a noise-like signal there is only a statistical likelihood that it contains power at a specific frequency, and over the whole spectrum some frequencies will have reasonable power content and some very little. The distribution will change with each recording of the signal. Hence single spectra of a noise-like signal are quite uneven and variable. If you average several spectra, however, it becomes much more likely that there will be reasonable power present at all frequencies and the averaged spectrum is much closer to the theoretical flat line. When working with noise-like signals, it is therefore very common to take several recordings and average their spectra to obtain useful results. Each individual recording is called an *epoch* and the average is referred to as an *ensemble* average. It is important to note that the spectra are averaged, not the raw data. The spectrum of each epoch needs to be calculated separately and the results averaged. Figure 6.26 shows four individual power spectra calculated on successive sections of the data set. They were calculated by taking a sixty-four-sample section of the data, applying a Hamming window and squaring the magnitude of the resultant spectrum. Some features such as the peak at 1.5 Hz are consistent from one spectrum to the next but there is considerable variation and it would be difficult to have much confidence in values derived from just one of the spectra. We need to calculate the ensemble average of the spectra to obtain useful results. To do this we chop the data up into epochs, calculate the power spectrum of each one and average the results. It is also usual to overlap the epochs to some extent but the actual choice of epoch length and overlap percentage is fairly arbitrary. The more epochs per recording the better because the variation in the ensemble spectrum is reduced

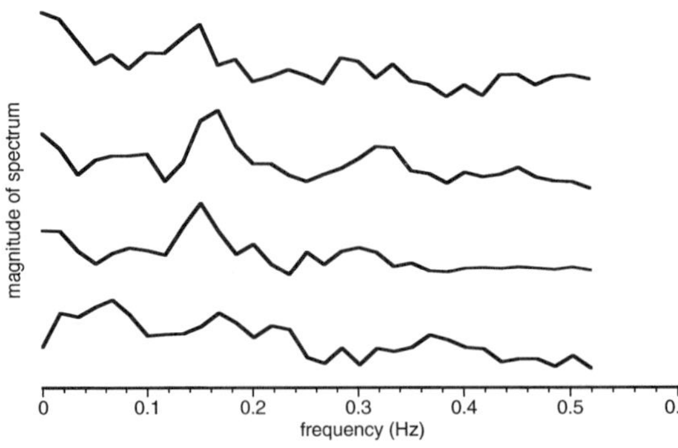

Figure 6.26 The magnitudes of four spectra calculated from successive sixty-four-sample sections of the data in Figure 6.25. There is considerable variation between the four spectra.

as the number of epochs is increased – the variation is inversely proportional to $\sqrt{\text{number of epochs}}$. Typically, 10–30 epochs are averaged to give a good spectrum for heart rate variability. The upper limit to the number of epochs is set by the required spectral precision because the bin width is 1/(epoch duration) and so short epochs will give poor spectral resolution. A slight twist when dealing with heart rate variability data is that the duration of each epoch in seconds is not known, since the data are sampled once per beat rather than on a regular time basis. The best we can do is to calculate the average R–R interval and substitute this for the sampling rate. In this case the mean R–R interval is 933 ms so the 'sampling rate' is 1.07 Hz. A sixty-four-sample epoch is 59.72 s long giving a spectral resolution of 0.017 Hz.

Figure 6.27 shows the ensemble-averaged power spectrum of the heart rate data. The epoch length was sixty-four samples and successive epochs overlapped by 50%, so eleven complete epochs were squeezed out of the data and the last sixteen samples were ignored. The power spectrum of each epoch was calculated as for Figure 6.26 and the mean of the eleven power spectra is shown in Figure 6.27. A prominent peak at 0.15 Hz is due to respiration and the respiratory rate can be calculated as $0.15 \times 60 = 9$ breaths/min. There is another peak at 0.33 Hz and considerable low-frequency energy. The overall shape of the heart rate variability spectrum has also been analysed from a different perspective. If the individual peaks in Figure 6.27 are ignored there is a general trend of decreasing power as the

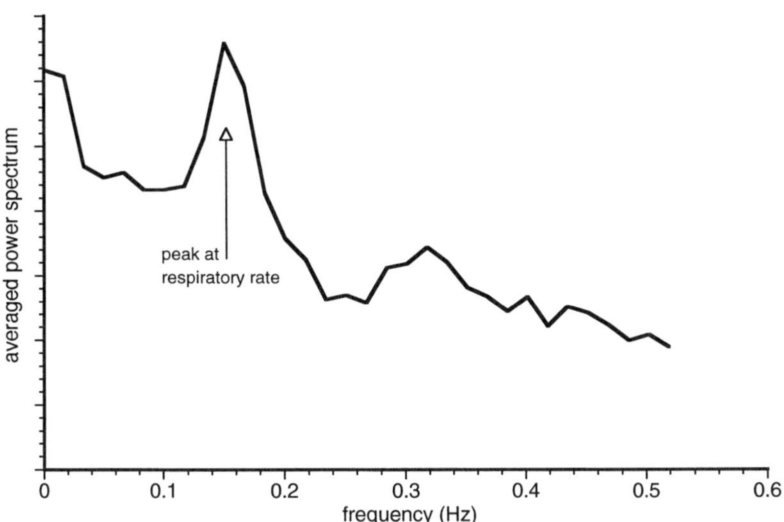

Figure 6.27 Ensemble average of the power spectra of eleven overlapping sixty-four-sample sections of the data in Figure 6.25.

frequency increases and at least two theories have been put forward to explain it. One is that it is flicker or $1/f$ noise which is widely found in natural processes. The other theory is that there is a chaotic component to the heart rate. Chaos is a nonlinear phenomenon that differs from random processes because a chaotic process is predictable on a short time scale, whereas a truly random process is unpredictable. The mathematical techniques used to analyse nonlinear phenomena (e.g. Barahona and Poon 1996) are complex and tend to need large data sets but have been used to search for chaotic processes in heart rate data.

EQUINE HOOF ANATOMY

Figure 6.28 shows some data from an anatomical study of the structure of a horse's hoof. The hoof wall is joined to the soft tissues and bone inside the foot by a series of interdigitations called laminae. The laminae transfer all the forces from the hoof to the inside of the foot and any abnormality of the laminae seriously affects the horse's athletic ability. In this study the investigator was interested in the orientation of the laminae with respect to the hoof wall and wanted to know how the pattern of orientation changed between feet and in different places in the same foot. Figure 6.28 shows the angle at which each lamina meets the hoof wall, measured using image analysis techniques on a digitized photograph of a cross-section of a hoof. There are about 500 laminae in each foot and they meet the hoof

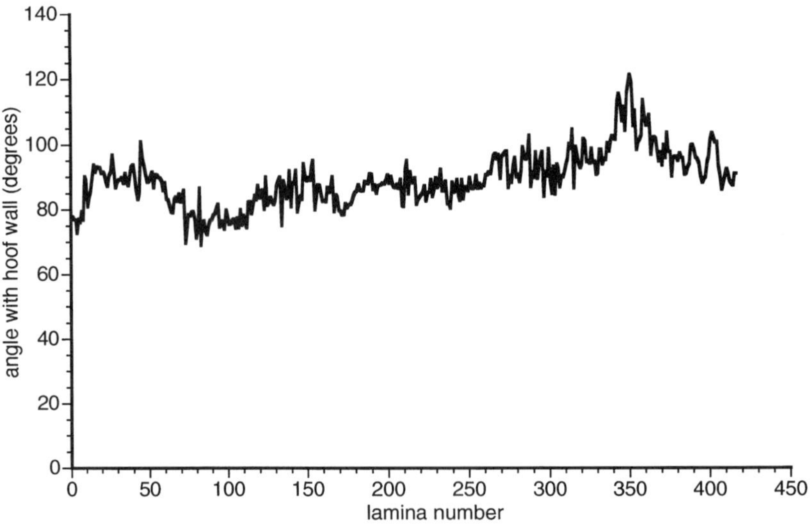

Figure 6.28 The orientation of the laminae in a horse's hoof with respect to the hoof wall.

wall at roughly 90°, but the actual angle varies slightly around the foot. We are interested in this variation and how it changes between feet.

Like the heart rate data in the previous example, the data in Figure 6.28 are 'auto-sampled' because each lamina has only one orientation and so we can record only one angle per lamina. Hence the results are automatically sampled once per lamina. If we were to perform spectral analysis on the data, the frequency scale would be in 'cycles per hoof' – an odd scale but perfectly valid. As in the heart rate example, we assume that the samples are evenly spaced even though there is some variation in interlaminar distance. The first impression of the data in Figure 6.28 is that there is probably an underlying pattern with three or four peaks in the whole data set, mixed in with a lot of high frequency noise. The latter is partly real and partly due to the limited resolution of the digitizing system used to image the foot. Our first task is to remove the DC offset by subtracting the mean value of the data set from each value because we are not interested in the absolute value of each angle (which is about 90°) but the variation in angle around the foot. Figure 6.29 shows the result. The next task is to remove the noise using a low-pass filter but it is not at all clear what corner frequency of filter should be used. To decide this the spectra of several digitized images were calculated using the DFT rather than the FFT, thus avoiding having to truncate the data sets or pad them to make the number of samples a power of 2.[4] Visual inspection of the magnitudes of the spectra indicated that there was no consistent pattern beyond the fifth bin.

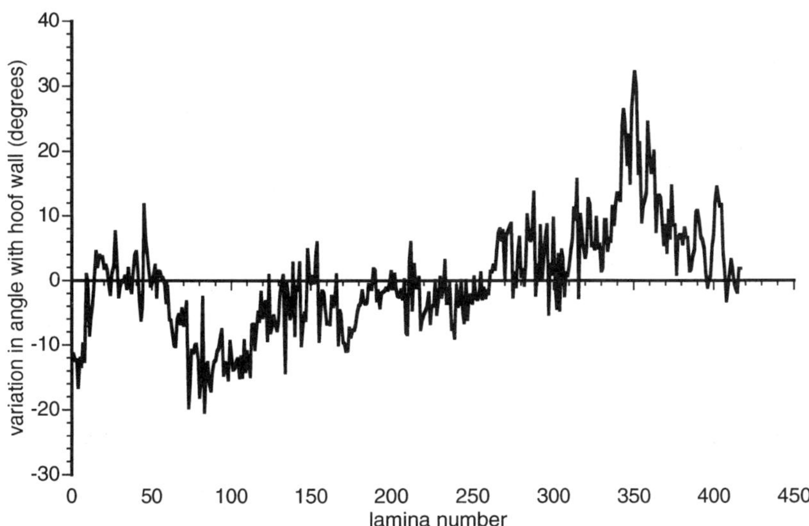

Figure 6.29 The data from Figure 6.28 with the DC offset removed. The vertical scale has been expanded.

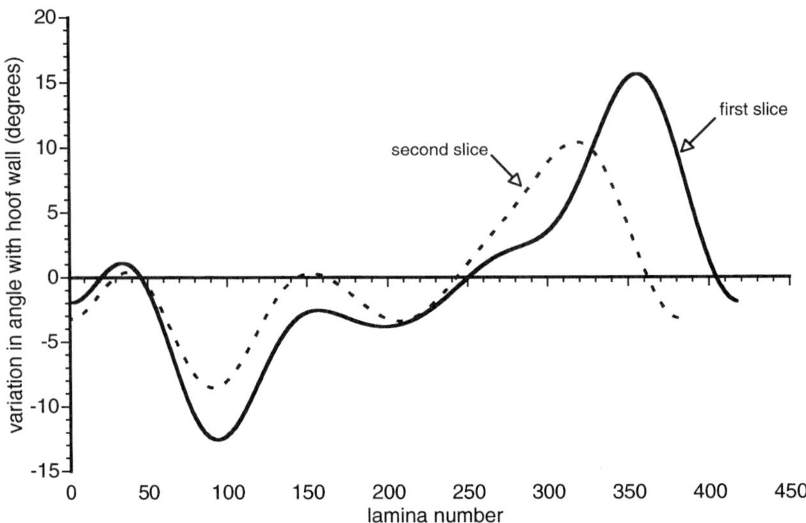

Figure 6.30 The data from Figure 6.29 after low-pass-filtering (solid line). Also shown is the laminar orientation pattern from the next slice of the foot (dashed line). Note that the pattern of variation of orientation between the two slices is similar but the number of laminae are different.

The data were then low-pass-filtered digitally with the corner frequency set to 4 cycles/hoof in a slightly unusual way. The Fourier transform of each data set was calculated, preserving all the values rather than just the first half of the spectrum. All points except for the first and last five were set to 0 (both magnitude and phase) and then the inverse transform was calculated. The result was a data set low-pass-filtered at 4 cycles/hoof and this is shown in Figure 6.30. Now the underlying pattern of angle variation can be seen much more clearly.

The trace labelled 'first slice' in Figure 6.30 shows the low-pass-filtered data from Figure 6.29. The dashed trace is the low-pass-filtered data from a second slice further down the hoof. A similar pattern of angle variation is visible in the second slice, and we now have to quantify how 'similar' the two slices are. Cross-correlation is a good way of expressing 'how much' of one signal is present in another and seems appropriate. But one problem is that the number of laminae varies from one slice to another and one foot to another: the first slice contains 417 laminae and the second one 381. The pattern, on the other hand, seems to be the same for the whole foot no matter how many laminae are present. If we cross-correlate two data sets with the same overall pattern but different numbers of samples we will get a lower correlation coefficient than we should because the cross-correlation process will line up the two traces point for point. We therefore need a way to preserve the pattern in the data but make all the data sets have the same number of

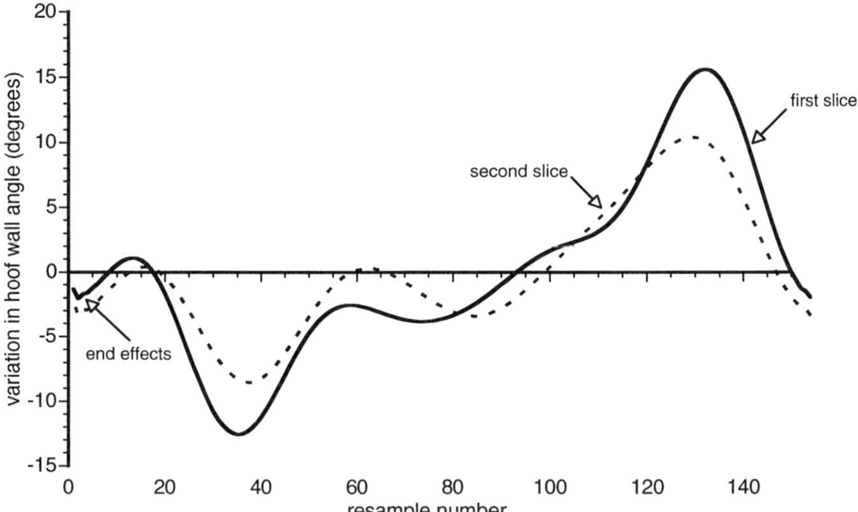

Figure 6.31 The two data sets from Figure 6.30 after resampling to 150 points each. The patterns on the two slices now line up.

points, which was done by resampling the data sets to a fixed number of points. In this study a value of 150 points was chosen because it was a little lower than the smallest number of laminae found in any foot. Before resampling to a lower rate the data should be low-pass-filtered, which in this case has already been done. Figure 6.31 shows the two traces from Figure 6.30 resampled to 150 points and it is clear that the pattern of variation has been preserved but there is some distortion at the ends of each data set. These 'end effects' are an artefact of the algorithm used to resample the data. Matlab was used to perform the resampling and it has a fairly sophisticated algorithm that models a short section of data around each point to obtain the best estimate of the value of the intermediate point. Because it needs several points to make this estimate the algorithm has trouble at the ends of the array where these points are not available.

Finally, a cross-correlation can be performed on the two filtered and resampled data sets. The cross-correlation was performed only with the two arrays completely aligned (zero lag) yielding a single number that represents how well the two arrays match each other. In this case the value was 0.954, which indicates that the two data sets were very similar. The cross-correlation was performed using Matlab, which has the advantage that it normalizes the result so that a perfect match is 1, regardless of the number of points or their actual values. The result of this type of cross-correlation is therefore a number between -1 and 1, with 0 indicating no similarity between the two data sets, 1 a perfect match and -1

indicating that the data sets are mirror images of each other. Although calculating the cross-correlation coefficient requires many successive signal-processing steps on the data, it yields a number that is an objective measurement of how similar the pattern of lamina angle variation is between successive slices.

NEUROPHYSIOLOGICAL RECORDING

Figure 6.32 shows data recorded from isolated brainstem tissue in an organ bath. The recording was made from the phrenic nerve and represents the output of the respiratory centre. The data were high-pass-filtered with a corner frequency of 100 Hz before digitizing. There is a constant background noise from the amplifier and superimposed on this are five bursts with much higher amplitude. The

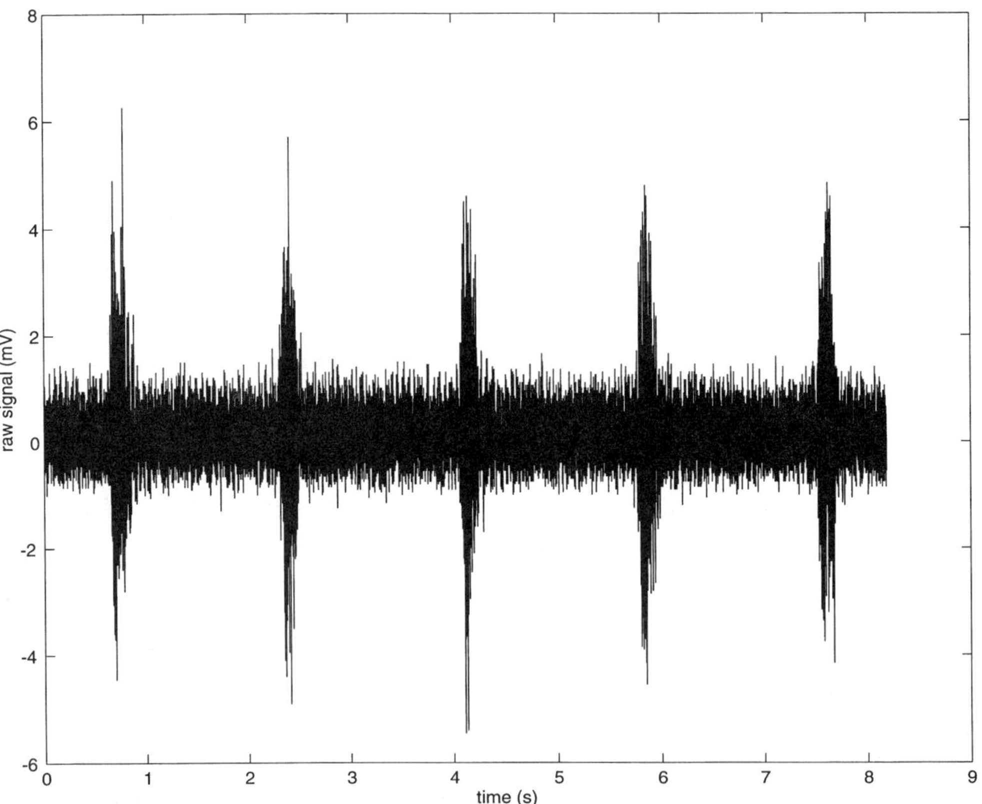

Figure 6.32 Amplified signal from the phrenic nerve of isolated brainstem tissue. The recording was made using a sampling rate of 4 kHz. There is a constant background noise, superimposed upon which are five bursts of neural activity. Each burst consists of hundreds of action potentials.

phrenic nerve supplies the diaphragm and the phasic bursts of activity would cause the diaphragm to contract in the intact animal. Recordings were made from an electrode hooked around the nerve. The nerve is made up from a bundle of motor neurone axons, and when an action potential travels down an axon, it generates a small spike in the recording electrode.

The first problem with recording action potentials like these is that the sampling rate has to be high. Each action potential is only 1.5 ms wide and thus the sampling rate has to be in the kHz range just to make sure that each action potential is recorded. The signal in Figure 6.32 was recorded at 4 kHz and thus even this short segment, containing five breaths, consists of 32,765 samples. Storing and processing these digitized signals is a problem and the recording system used for this example has a digital audio tape (DAT) system for bulk storage.

The action potentials in the nerve consist of narrow spikes with fairly constant amplitude. One way of analysing the data would be to treat the spikes as digital signals. The height of an action potential does not carry any information because it is constant: all the information is in the timing of the pulses. The amplified action potentials could therefore be cleaned up with a threshold circuit and sent to a digital counter. Unfortunately, this approach can only be used in small nerves where there are fewer axons because each action potential has to occur separately. If two axons fire within a few ms of each other the two action potentials will summate to form a single, larger pulse. The digital counter will only see a single pulse and thus the count rate will be inaccurate. The digital approach works when there are only a few fibres in the nerve and the probability of two action potentials occurring together is very low. In this example the nerve contains thousands of axons and the digital approach is no good.

The basic information that we need to extract from this signal are the size and frequency of the bursts. We are not very concerned with the individual action potentials, we need to know how strong each burst is (which represents the amplitude of respiration) and their frequency (which is the respiratory rate). The basic approach used is to integrate the signal. Each action potential is a narrow spike with fairly constant width and amplitude. It therefore has a fairly constant area when it is integrated. If two action potentials occur close to each other their amplitudes will summate into a single, larger pulse whose integral will be twice that of a single pulse. This is very useful because if we integrate the signal the result will reflect the number of action potentials that occurred, regardless of whether they were superimposed or not.

Integrating the raw signal in Figure 6.32 will not be very rewarding, however. The signal is quite symmetrical about the 0 V point and so its integral will be zero, because the positive and negative areas are similar. The bipolar nature of the signal is a consequence both of the action potential itself (which has a depolarization and repolarization stage with opposite polarities) and the recording system. Often, a pair of recording electrodes are placed on the nerve a few mm apart. The two electrodes are connected to the $+$ and $-$ inputs of a differential amplifier. This reduces noise but it makes the signal bipolar because each action potential passes under both electrodes. As it passes under the $+$ electrode, the amplifier records a spike in one direction. The action potential then passes under the $-$ electrode and an identical spike of opposite polarity is recorded. The signal is centred around $0\,V$ because it was high-pass-filtered before being digitized. The high-pass filter removed any DC offset and the signal settled so that its mean value was $0\,V$.

The first task, therefore, is to convert the bipolar signal into something that has a meaningful integral. This is done by rectifying the signal.[5] There are two ways of rectifying a signal. Half-wave rectification simply throws away half of the signal. When the signal is positive it is unaffected, and when it is negative it is set to zero. Half-wave rectification is simpler to implement electronically but is inefficient because half of the signal is thrown away. Full-wave rectification preserves all of the signal's energy. As before, the positive half of the signal is left alone. The negative half is converted to an equivalent positive signal, that is, its sign is reversed. Both forms of rectification are very easy to implement digitally. Half-wave rectification is performed by looking at the sign of each sample. If it is positive it is left alone, and if it is negative it is set to zero. Full-wave rectification is even simpler because the sign of every sample is set to +ve. Most languages have an 'absolute value' operation to perform this task. Figure 6.33 shows the result of full wave rectification of the signal.

Now that the signal is rectified, it can be integrated. Figure 6.34 shows the integrated signal. Integration was performed using the trapezium rule and ignoring the factor of 0.5 for the first and last samples. Each sample was multiplied by its width δt, which is $1/4{,}000\,s$ since the sampling rate is $4\,kHz$, and the samples were then summed. The result is both good news and bad news. The good news is that the spiky signal has been converted into a much smoother waveform and the overall size and timing of each breath are clear. The bad news is that the constant noise in the signal is also integrated and produces a steadily rising baseline that makes it difficult to perform further measurements. Subtracting a constant offset from the signal before integration is not a good solution because, no matter how hard you try, the integrator will drift in one direction or another. Also, the amplitude of the noise will not be constant and the offset will have to be continuously adjusted. The solution to the problem is to use a *leaky integrator*. The term 'leaky' comes from the electrical circuit used to perform this function. Figure 6.35 shows

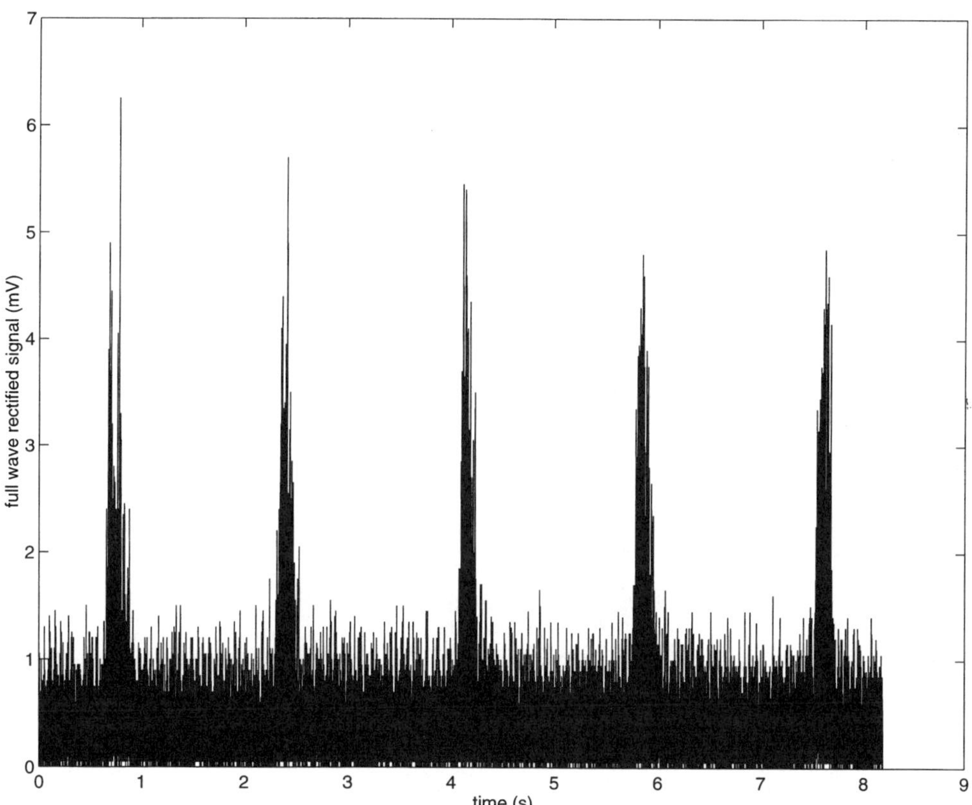

Figure 6.33 The signal from Figure 6.32 after full-wave rectification. This was accomplished by setting the sign of every sample to positive.

a classic integrator circuit. Without R_2 the circuit is a perfect integrator, the output being the integral of the input:

$$\text{output} = \frac{-1}{R_1 C} \int \text{input } dt$$

where $R_1 C$ is the time constant of the integrator and is a scaling factor that determines how quickly the output changes. If a constant $-1\,\text{V}$ signal is applied to the input of the integrator the output rises at $1/R_1 C$ volts per second.

If R_2 is added the integrator becomes leaky, because R_2 'leaks' charge out of the capacitor. The leak affects the low-frequency operation of the circuit. A perfect integrator has an infinite gain at DC, so if a constant voltage is applied to the input the output keeps on changing. A leaky integrator has a high but fixed gain of $-R_2/R_1$ at DC (R_2 is always much larger than R_1). As the frequency increases,

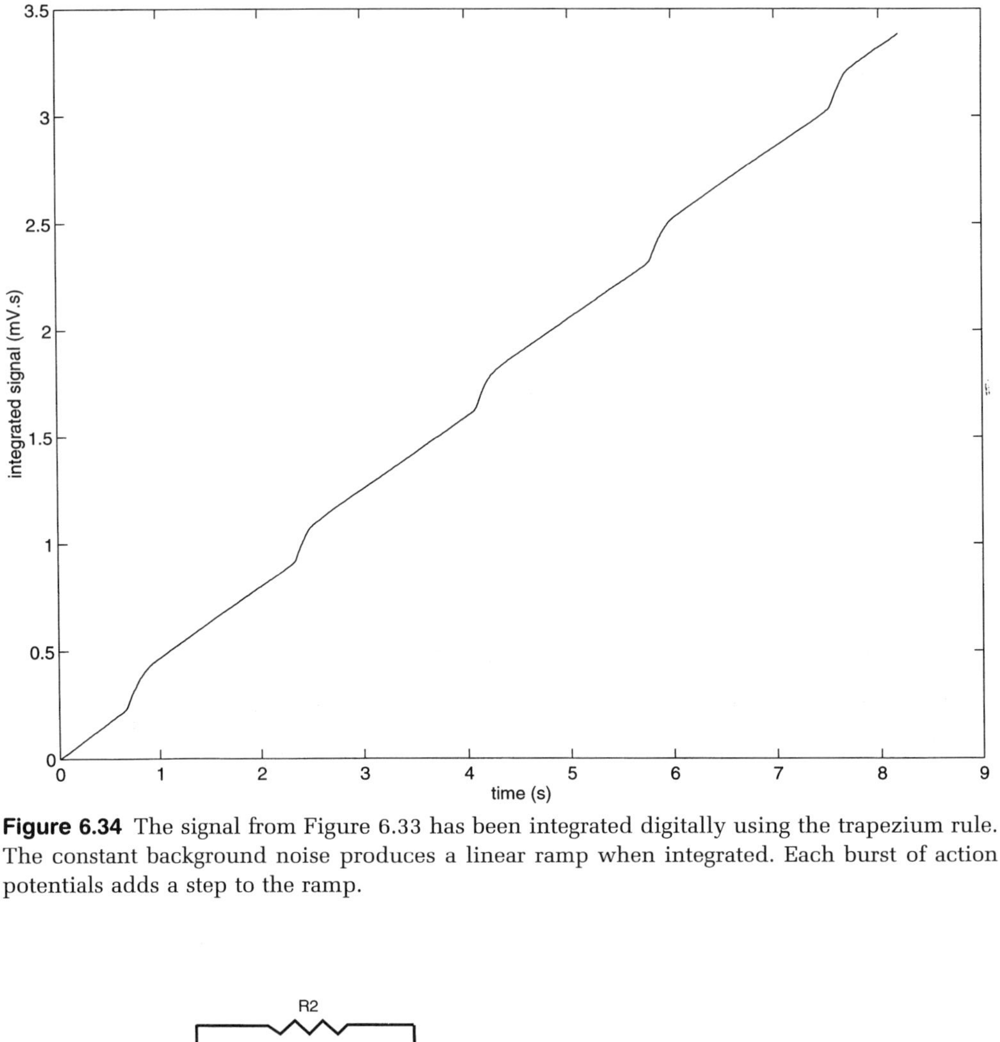

Figure 6.34 The signal from Figure 6.33 has been integrated digitally using the trapezium rule. The constant background noise produces a linear ramp when integrated. Each burst of action potentials adds a step to the ramp.

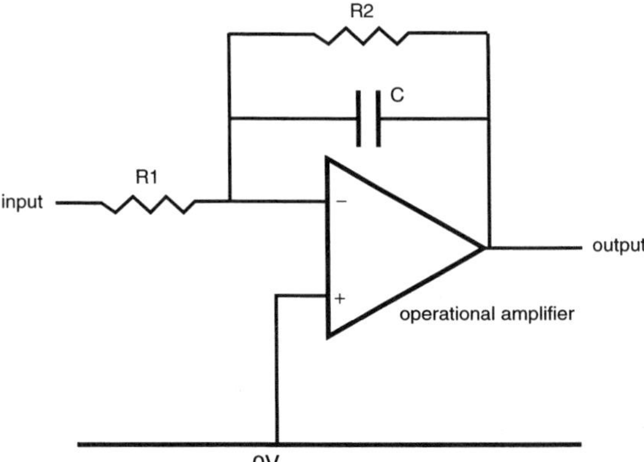

Figure 6.35 The basic circuit diagram for a leaky integrator. Without R_2 the circuit is a standard integrator whose output is the integral of the input: output $=(-1/R_1 C)\int$ input dt. With R_2 the integrator becomes leaky. At high frequencies $(f \gg 1/R_2 C)$ the circuit acts as an integrator. At low frequencies $(f \ll 1/R_2 C)$ the circuit acts as an amplifier with a gain of $-R_2/R_1$.

the leaky integrator acts more and more like a perfect integrator. The crossover frequency from leaky to normal operation is set by R_2 and C: the integrator is leaky for frequencies well below $1/R_2C$ and normal for frequencies well above $1/R_2C$.

A leaky integrator has the useful property that it gradually 'forgets' the effects of previous inputs. Instead of the output representing the perfect integral of all previous inputs, the output is weighted and recent inputs have a greater influence on the output than do older ones. Leaky integrators do not drift like perfect integrators. Integrator drift is caused by small offsets building up as integration proceeds. A leaky integrator gradually forgets past inputs and thus small offsets are not able to build up.

How do we create the digital equivalent of a leaky integrator? A perfect integrator using the trapezium rule calculates the sum of all the samples, multiplied by their width. If we call the integrated signal y_n ($n = 1$ to the total number of samples), and the input signal x_n, then

$$y_n = \delta t(x_1 + x_2 + x_3 + \cdots + x_n)$$

where $\delta t = 1/(\text{sampling rate})$, and the factor of 0.5 has been ignored for the first and last samples.

One way to create a leak is to weight each sample before summing. If we give new samples a weight of 1, and older samples a weight that gradually reduces from 1 to 0 as the sample gets older, we will get the desired result. So our leaky digital integration is

$$y_n = \delta t(w_n x_1 + w_{n-1} x_2 + w_{n-1} x_3 + \cdots + w_1 x_n)$$

where w_1, w_2, w_3,... are weights that decrease gradually from $w_1 = 1$ towards 0, with w_n being the smallest weight since x_1 is the oldest sample. This will achieve the desired result but it is computationally very inefficient. For each new output sample a new set of weights has to be calculated and applied to the input samples before summation.

Before moving on to a more efficient algorithm, consider for a moment how the weights are calculated. Any function that decreases smoothly from 1 towards 0 will work and in the digital version of the leaky integrator we are not restricted to one particular function. However, it is interesting to see how the weights are 'calculated' in the analogue version in Figure 6.35. Suppose that the integrator has a switch so that R_2 can be switched in and out of the circuit. Start with R_2 out of the circuit. Apply a voltage to the integrator and allow the output to rise to some value, for example, 2 V. Change the input to 0 V. The output will remain at 2 V because the integrator is perfect. Now switch in R_2. The voltage across the capacitor, and thus the output, will fall off exponentially towards 0 according to the formula

$$\text{output} = 2e^{-t/R_2 C}$$

where t is the time after the switch is closed and the value 2 is the output voltage when the switch was closed. Thus the 'weights' decrease exponentially with time. We can do the same in our digital system by calculating the weights as follows:

$$w_n = e^{-n\delta t/\tau}$$

where $\delta t = 1/(\text{sampling rate})$ and n is the sample number ($1 = $ most recent). The parameter τ determines how rapidly the weights decrease towards 0 and is equivalent to $R_2 C$ in the analogue system.

A more efficient algorithm for the digital leaky integrator is needed if the idea is to be of any use. Consider the perfect integrator once again:

$$y_n = \delta t(x_1 + x_2 + x_3 + \cdots + x_n)$$

The first output sample, y_1, is given by $y_1 = \delta t\, x_1$. The second sample is given by

$$y_2 = \delta t\, x_1 + \delta t\, x_2 = y_1 + \delta t x_2$$

Similarly,

$$y_3 = \delta t\, x_1 + \delta t\, x_2 + \delta t\, x_3 = y_2 + \delta t\, x_3$$

and so on. In general, a new output sample $y_n = y_{n-1} + \delta t\, x_n$. This 'difference equation' shows that a new output sample can be generated from the previous output sample plus the contribution from the current input sample. Besides being an efficient way of calculating the perfect integral, we can use this format to generate a leaky integrator.

Suppose we now weight each previous output by α before adding the contribution from the current sample, where α is a constant value. How does this affect the integral? We now have

$$y_n = \alpha y_{n-1} + \delta t\, x_n$$

However, y_n is similarly made up from $\alpha\, y_{n-2}$ and $\delta t\, x_{n-1}$. Expanding this for a few output samples, we get

$$y_n = \alpha y_{n-1} + \delta t\, x_n$$
$$= \alpha^2 y_{n-2} + \alpha y_{n-1} + \delta t\, x_n$$
$$= \alpha^3 y_{n-3} + \alpha^2 y_{n-2} + \alpha y_{n-1} + \delta t\, x_n$$

Notice that as the output samples used to make up y_n get older they are multiplied by successively higher powers of α. If we set α to be a number between 0 and 1 we will achieve our goal. Higher powers of α will gradually decrease towards 0, and thus the output y_n will be made up from the weighted sum of the contributions of the individual input samples, with the weights becoming smaller as the samples become older. Furthermore, the weights will 'automatically' update with each new output. The algorithm is very efficient because we only need two multiplications and one sum to calculate a new output sample.

The quantity α is called a 'forgetting function' and its value must be between 0 and 1. The closer to 0, the faster the integrator 'forgets' (or the 'leakier' it becomes). Typical values of α are just below 1, for example, 0.99. For many applications a value of α can be picked empirically to give the required forgetting rate. On the other hand, if we want to emulate the analogue integrator as closely as possible we can calculate the required value of α from R_2 and C using the following argument. General exponential decay is described by the equation

$$y_t = e^{-kt}$$

Make the substitution $a = 1/e$. Then

$$y_t = e^{-kt} = (1/e)^{kt} = a^{kt}$$

$k = 1/R_2 C$ for the leaky integrator, so $y_t = e^{-t/R_2 C} = a^{t/R_2 C}$. In the digital equivalent, t is replaced by $n\delta t$ so the weights become

$$w_1 = a^{\delta t/R_2 C}$$
$$w_2 = a^{2\delta t/R_2 C} = (a^{\delta t/R_2 C})^2$$
$$w_3 = a^{3\delta t/R_2 C} = (a^{\delta t/R_2 C})^3, \quad \text{etc.}$$

Thus $\alpha = a^{\delta t/R_2 C} = e^{-\delta t/R_2 C}$ (since $a = 1/e$). So if we want to create the digital equivalent of the leaky integrator in Figure 6.35 we use a forgetting function α where $\alpha = e^{-\delta t/R_2 C}$.

Figures 6.36 and 6.37 show the digital leaky integrator in action. Figure 6.36 is the integrated signal using an empirical forgetting function of $\alpha = 0.99$. Figure 6.37 mimics the analogue leaky integrator in Figure 6.35 with the leak rate $R_2 C$ set to 0.05 s. Also, the scaling function $R_1 C$ was set to 20 ms, which accounts for the

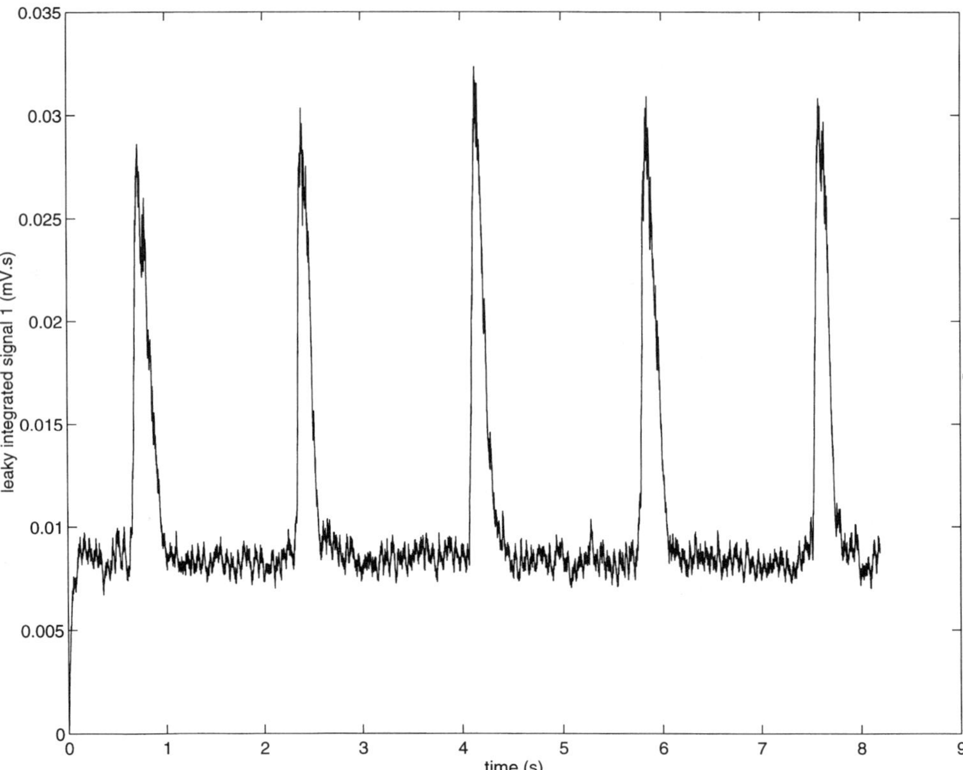

Figure 6.36 The signal from Figure 6.33 after digital leaky integration, using a forgetting function of 0.99. Instead of the noise integrating to a linear ramp it forms a DC offset. The bursts of action potentials, which consist of many narrow spikes, integrate into a broad peak whose shape follows the outline or 'envelope' of the bursts.

different y-axis scales between the two graphs. The peaks are a little broader and the noise a little lower in Figure 6.37 compared to Figure 6.36. This is because the forgetting function in Figure 6.37 was 0.995 (α), compared to 0.99 in Figure 6.36. The larger forgetting function produces a slower leak, which in turn increases the low-frequency gain of the leaky integrator (see below). Figure 6.38 shows what happens when half-wave rectification is used rather than full-wave: the loss of signal gives a poorer signal-to-noise ratio.

The leaky integrated signal (Figure 6.37) has a lot of noise compared to the perfectly integrated signal (Figure 6.34), which is very smooth. The noise is a consequence of the leak, which reduces the gain at low frequencies. A perfect integrator has a very large low-frequency gain (gain is proportional to 1/frequency), whereas a leaky integrator has a fixed low-frequency gain. The perfect integrator therefore emphasizes the very low frequency components of the signal,

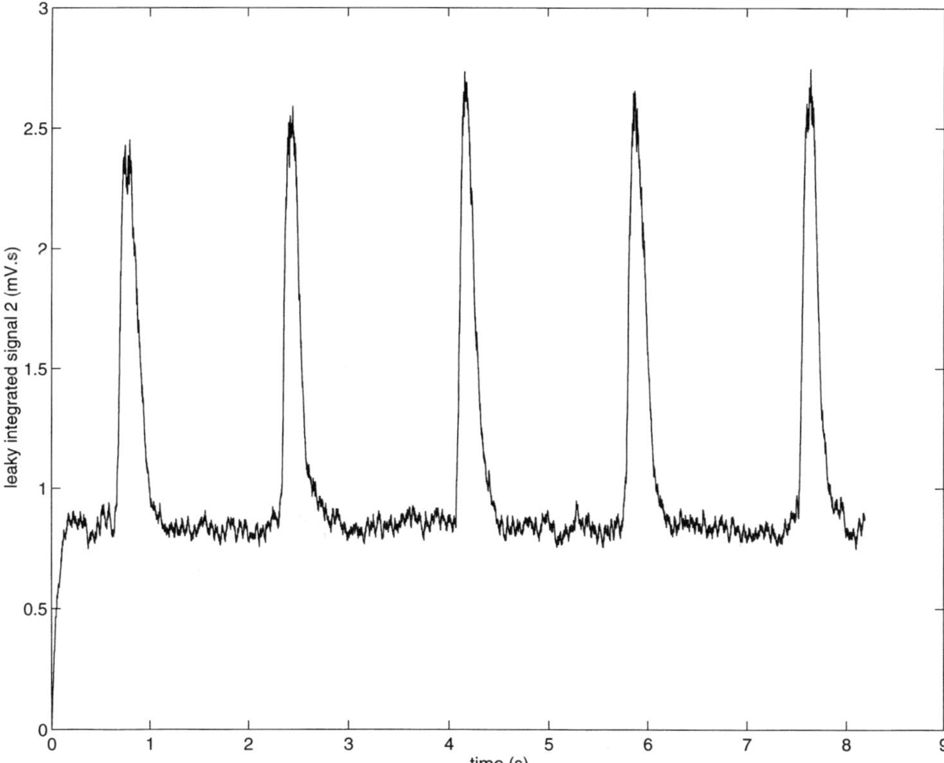

Figure 6.37 The signal from Figure 6.33 after digital leaky integration using a forgetting function of 0.995. Since the forgetting function is closer to 1 the integrator is less leaky than the one in Figure 6.36. This results in broader peaks and less high-frequency noise.

for example, the outline of the peaks. The leaky integrator will pass all signals with frequencies below $1/R_2C$ (the frequency at which the leak becomes important) equally well. In this signal there is a considerable amount of noise with frequency below $1/R_2C$, which appears in the leaky integrated signal. Another minor point is that the leak makes it difficult to determine the units of the leaky integrated signal. A perfectly integrated signal has units of V s. A leaky integrated signal acts as an amplifier at low frequencies, and thus has units of V at low frequency and V s at high frequencies. In this example we will call the units of the integrated signals, however they are obtained, V s simply for clarity.

The apparatus used to collect the signal had a built-in analogue leaky integrator. The integrator had its time constant (R_1C in Figure 6.35) set to 20 ms but the time constant of the leak (R_2C) was not specified. The low-pass-filtered signal from the amplifier was also sent to the analogue integrator. The output of the integrator

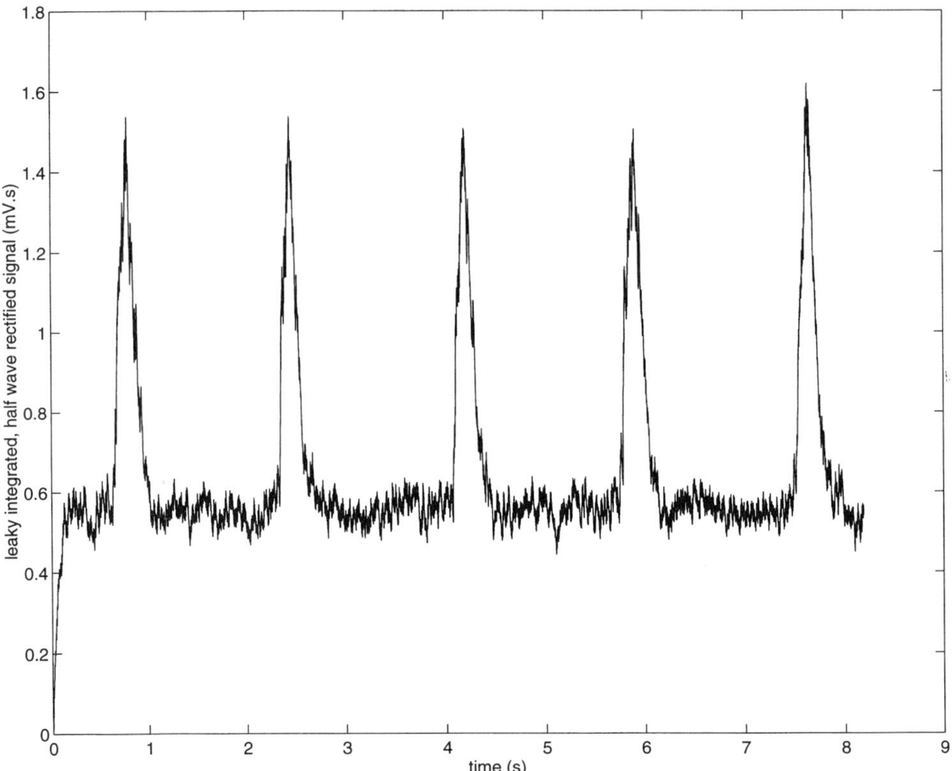

Figure 6.38 The signal in Figure 6.32 was half-wave, rather than full-wave, rectified before integration. The loss of information results in a noisier signal compared to the full-wave rectified version in Figure 6.37.

was digitized and it is interesting to compare the digitally processed signal with its analogue counterpart. Figure 6.39 shows the signal after analogue rectification, leaky integration, and digitization. The shapes of the peaks are almost identical to the digital version in Figure 6.36, indicating that the forgetting function was approximately 0.99. This corresponds to a leak time constant $(R_2 C)$ of 25 ms. However, the sampled analogue signal clearly suffers from poor resolution. The quantization steps are very obvious and are about 5 mV in amplitude. Without knowing anything about the data acquisition system we can guess that it is a twelve-bit (by far the most common) digitizer with an input range of ± 10 V, because this has a step size of 4.88 mV.

There is more processing that we can do on the digitally integrated signal. Firstly, it would be nice to cut down the number of samples. Now that the signal has been integrated we can low-pass-filter it because the original action potential

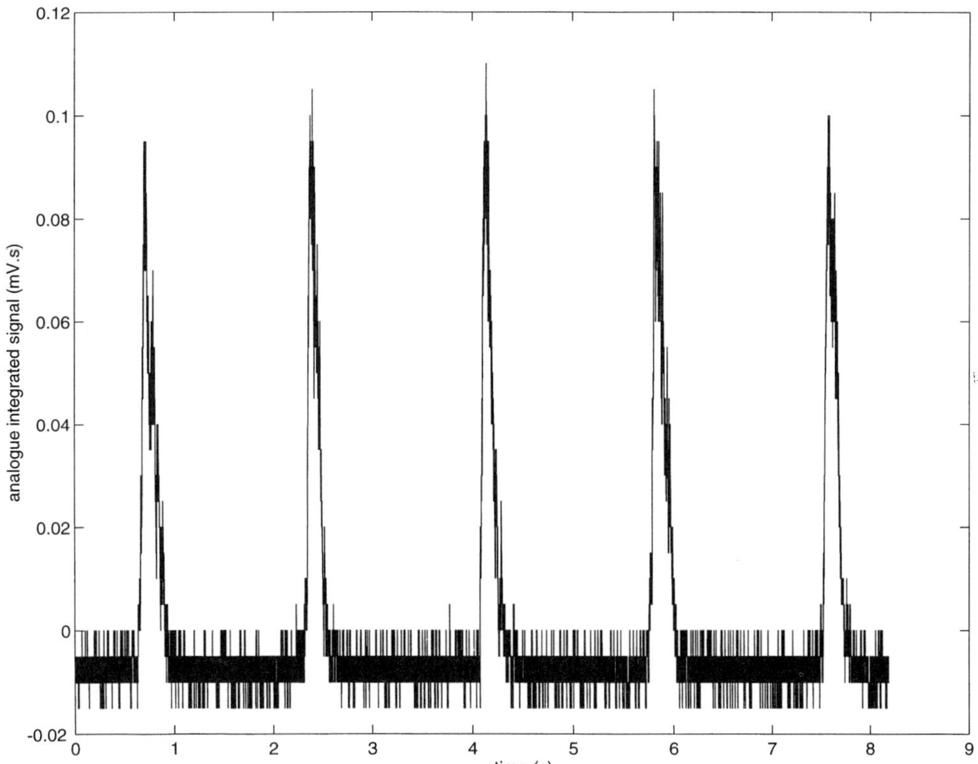

Figure 6.39 The signal from the nerve was also rectified and integrated using analogue circuitry before being digitized. The shapes of the peaks are almost identical to the digitally processed version in Figure 6.36. The amplitude of the signal was too low for the A/D convertor resulting in large quantization errors.

spikes have been integrated into broad peaks with much lower frequency content. Figure 6.40 shows the signal after low-pass-filtering with a sixth-order Butterworth filter with the corner frequency set to 20 Hz. The information in the peaks has been preserved but the high-frequency noise has gone. Now that the data are low-pass-filtered we can resample the signal at a much lower rate. The corner frequency of the filter was 20 Hz, so there will not be much signal present above 40–50 Hz (the sixth-order filter has a roll-off rate of 36 dB/octave). Figure 6.41 shows the signal resampled at 100 Hz, 1/40th of its original rate.

Finally, we are ready to make some measurements on our data. We are interested in the size of the peaks and their positions. The first task is to remove the baseline offset, which is due to amplifier noise. We cannot just calculate the average of the whole signal and subtract it because the average value is higher than the baseline noise. What we need is the average value of the noise in between

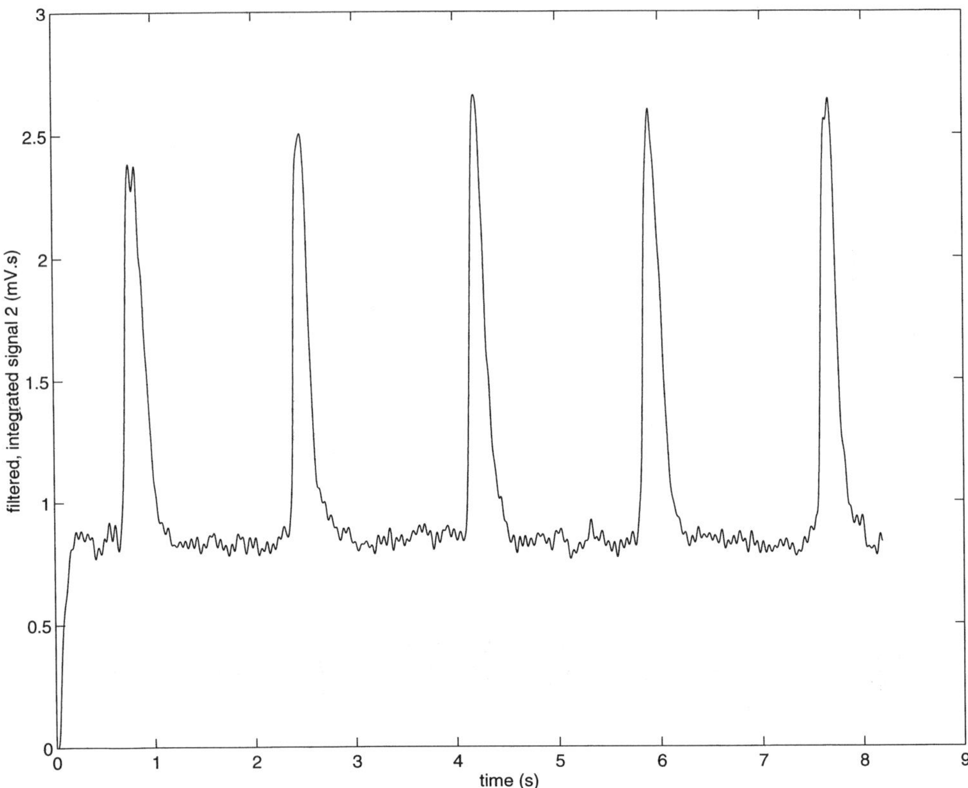

Figure 6.40 The signal from Figure 6.37 after low-pass-filtering with a sixth-order Butterworth filter. The corner frequency was 20 Hz.

the peaks. A peak-detection algorithm will find the start, end, and maximum values of the peaks. The hysteresis band needs to be set to just above the noise level, for example, between 1.0 and 1.3 mV s (see Figure 6.41). The average of the signal, omitting the portions between the start and end of the peaks, is then calculated (0.83 mV s) and subtracted from the whole signal. The heights of the peaks can now be measured (2.383, 2.505, 2.660, 2.600, and 2.642 mV s). The peaks occur at 0.77, 2.47, 4.20, 5.90, and 7.67 s after the start of the trace. The interpeak intervals are therefore 1.70, 1.73, 1.70, and 1.77 s with a mean value of 1.73 s, making the respiratory rate 34.7 breaths/minute.

The heights of the peaks are rather variable because they occur when a few neurones fire very close in time to each other, an event with a low probability. It seems a pity to have a peak made up from thousands of individual action potentials, characterized by just a few of those action potentials. A better metric would

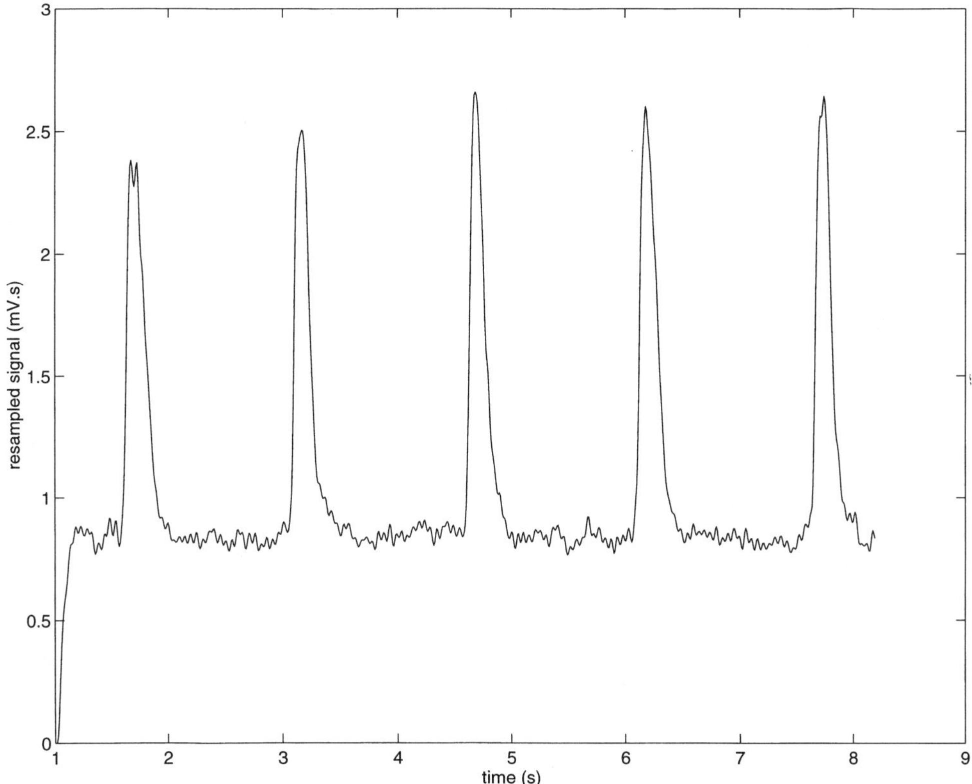

Figure 6.41 The low-pass-filtered signal was resampled at 100 Hz, 1/40th of the original rate. The reduction in the number of samples makes it much easier to perform further processing on the signal.

be to include all the action potentials. One way to do this is to calculate the area of the peak. This also has the advantage that it reflects the overall respiratory effort, whereas the peak height reflects the maximum firing rate. The areas of the peaks are easily calculated (using the trapezium rule) because we have already found the start and end of the peaks. In this example the areas of the peaks are 0.329, 0.311. 0.322, 0.338, and 0.338 mV s^2. The rather unusual units reflect the fact that the data have been integrated twice.

The digital signal processing in this example was performed in Matlab. The algorithms used may be of interest to some readers and the code is given here. The data are in files raw.dat and an_integ.dat and are in ASCII format. Although it is good for demonstration purposes, using Matlab to perform the calculations is inefficient. A better language would be LabView, which could process the data in real time on a fast computer.

```
% Analyse neurophysiological data recorded from
% isolated tissue.

% load in the raw data and the analogue integrated signal
load raw.dat
load an_integ.dat

% sampling rate is 4 kHz, dt is the sampling interval
dt=1/4000;

% half wave and full wave rectify the raw data
rect = (raw >= 0) .* raw;
fwrect = abs(raw);

% basic integration using the trapezium rule
integ=dt*cumsum(rect);
fw_integ=dt*cumsum(fwrect);

% now apply a leaky integrator using a forgetting function alpha
alpha=0.99;

% dimension the vector first and set first point,
% then integrate the rest of the signal
leaky1=fwrect;
leaky1(1)=fwrect(1)*dt;
for n = 2:length(fwrect)
     leaky1(n) = leaky1(n-1)*alpha + fwrect(n)*dt;
end

% the time constant for the leak = R2C seconds
% so the forgetting function is 1/exp(dt/R2C)
% Also, the time constant for the main integrator was
% 20ms so the integral is scaled by 1/20ms
% use the full wave rectified data
R2C = 0.05;
tc=0.02;
a=exp(-dt/R2C);
leaky2=fwrect;
leaky2(1)=fwrect(1)*dt/tc;
for n = 2:length(fwrect)
     leaky2(n) = leaky2(n-1)*a + fwrect(n)*dt/tc;
end
```

```matlab
% Do the same using the half wave rectified signal
hw_leaky2=rect;
hw_leaky2(1)=rect(1)*dt/tc;
for n = 2:length(rect)
    hw_leaky2(n) = hw_leaky2(n-1)*a + rect(n)*dt/tc;
end

% now lowpass filter the data using a 6th order Butterworth
% IIR filter. The cutoff frequency is fc (Hz). Remember that the
% sampling frequency is 4kHz so the Nyquist frequency is
% 2000 Hz. Matlab scales its frequencies to the Nyquist frequency.
fc = 20;
[d,c] = butter (6, fc*2*dt);
lp_leaky2=filter(d,c,leaky2);

% resample the data to make them more manageable
% fc = 20Hz so resampling at 100Hz is reasonable
% Matlab has the decimate function for this but
% decimate also applies a lp filter. To see the raw process,
% do it the hard way.
k=1;
for n = 1: 4000/100 :length(fwrect)
    resamp(k) = lp_leaky2(n);
    k=k+1;
end

% plot the results. Generate x scale for the original
% data sets and the resampled one.
x=[1:length(raw)]*dt;
resamp_x=linspace(1, length(raw)*dt, k-1);

% plot the results, pausing when each one is drawn. Press a key
% to see the next graph.
plot(x, raw); xlabel('time (s)'); ylabel('raw signal (mV)'); pause;
plot(x, fwrect); xlabel('time (s)'); ylabel('full wave rectified
signal (mV)'); pause;
plot(x, fw_integ); xlabel('time (s)'); ylabel('integrated signal
(mV.s)'); pause;
plot(x, leaky1); xlabel('time (s)'); ylabel('leaky integrated signal
1 (mV.s)'); pause;
```

```
plot(x, leaky2); xlabel('time (s)'); ylabel('leaky integrated signal
2 (mV.s)'); pause;
plot(x, hw_leaky2); xlabel('time (s)'); ylabel('leaky integrated,
half wave rectified signal (mV.s)'); pause;
plot(x, lp_leaky2); xlabel('time (s)'); ylabel('filtered, integrated
signal 2 (mV.s)'); pause;
plot(resamp_x,resamp); xlabel('time (s)'); ylabel('resampled signal
(mV.s)'); pause;
plot(x, an_integ); xlabel('time (s)'); ylabel('analogue integrated
signal (mV.s)');

% save the resampled data in ASCII format so that
% we can measure the heights of the peaks
save  resampled.txt resamp /ascii;
```

Suppliers of Data Acquisition/Analysis Hardware and Software and Electronic Components

ADInstruments Pty Ltd.
Unit 6, 4 Gladstone Road
Castle Hill
NSW 2154
Australia
+61 (2) 9899 5455

www.adinstruments.com
ADInstruments make a range of amplifiers and data-recording systems that run on both Mac and PC platforms. They specialize in life science systems and make isolated amplifiers for use on human subjects.

Allied Electronics Inc.
7410 Pebble Drive
Fort Worth
TX 76118
1 800 433 5700

www.allied.avnet.com
A useful supplier of electronic components. Good range of parts and small orders accepted.

Axon Instruments Inc.
1101 Chess Drive
Foster City
CA 94404
USA

Data acquisition and analysis hardware and software

Biopac Systems
5960 Mandarin Drive
Goleta
CA 93117
USA

Data acquisition and analysis hardware and software

Dataq Intruments Inc.
150 Springside Drive, Suite B220
Akron
Ohio 44333
USA

Data acquisition and analysis hardware and software

Datel Inc.
11 Cabot Boulevard
Mansfield
MA 02048-1194
USA

A/D and D/A hardware

DeltaPoint Inc.
2 Harris Court, Suite B-1
Monterey
CA 93940
USA

DeltaGraph graph plotting program

Digi-Key Co.
701 Brooks Avenue South
Thief River Falls
MN 56701-0677
1 800 344 4539

www.digikey.com
Another useful supplier of electronic components in the USA.
Fast service and small orders accepted.

GW Instruments
35 Medford Street
Somerville
MA 02143
USA

Data acquisition and analysis hardware and software

Intelligent Instrumentation
6550 South Bay Colony Drive, MS130
Tuscon
AZ 85706
USA

Data acquisition and analysis hardware and software

IOtech Inc.
25971 Cannon Road
Cleveland
OH 44146
USA

Data acquisition and analysis hardware and software

Jandel Scientific Software
2591 Kerner Boulevard
San Rafael
CA 94901
USA

Wide range of data analysis programs

Keithly Metrabyte
440 Myles Standish Boulevard
Taunton
MA 02780
USA

Data acquisition, signal processing and i/o hardware. Data acquisition and analysis software.

MathSoft Inc.
101 Main Street
Cambridge
MA 02142
USA

Mathematical, data analysis and graphing software packages

The Mathworks
Cochituate Place
24 Prime Park Way
Natick
MA 01760
USA

'Matlab' signal-processing software

Microcal Software Inc.
One Roundhouse Plaza
Northampton
MA 01060
USA

Data acquisition and analysis hardware and software

MicroMath Scientific Software
2469 East Fort Union Boulevard, Suite 200
P.O. Box 21550
Salt Lake City
UT 84121
USA

'Scientist' data analysis and mathematical software

Microsoft Corporation
One Microsoft Way
Redmond
WA 98052
USA

'Excel' spreadsheet program and Windows operating system

Microstar Laboratories
2265 116th Avenue NE
Bellevue
WA 98004
USA

> *Data acquisition and analysis hardware and software*

National Instruments
6504 Bridge Point Parkway
Austin
TX 78730
USA

> *Data acquisition hardware, 'LabView', 'LabWindows', and 'HiQ'*
> *software*

PC Science
16 Moraine Road
Morris Plains
NJ 07950
USA

> *Data acquisition and analysis hardware and software*

Po-Ne-Mah Inc.
883 Hopmeadow Street
Simsbury
CT 06070
USA

> *Data acquisition hardware and software*

Raytech Instruments Inc.
2735 West 31st Avenue
Vancouver
BC V6L 1Z9
Canada

> *Data acquisition hardware and software*

Remote Measurement Systems Inc.
2633 Eastlake Avenue East, Suite 200
Seattle
WA 98102
USA

Data acquisition hardware and software for low-sampling-rate environmental applications

Signal Centre
Broughton Grange
Headlands
Kettering
Northants NN15 6XA
UK

Data acquisition and analysis software

Spyglass Inc.
1800 Woodfield Drive
Savoy
IL 61874
USA

'Spyglass' data analysis/presentation software for 3-D work and large data sets

Strawberry Tree
160 South Wolfe Road
Sunnyvale
CA 94086
USA

Data acquisition and analysis hardware and software

WaveMetrics
P.O. Box 2088
Lake Oswego
OR 97035
USA

'Igor' data analysis program

Notes

Chapter 1

1. Assuming that a twelve-bit ADC is used, each sample occupies two bytes and data are stored in binary form. This is the most likely scenario.

2. The calibration graph can also be drawn with voltage on the vertical axis and temperature on the horizontal axis. The line is then

$$\text{voltage} = (\text{scale} \times \text{temperature}) + \text{offset}$$

   The values of the scale and offset are different in the two methods. It does not matter which method is used so long as consistency is maintained. If the scale and offset values might be used in other calculations it is important to state explicitly (using the equation of the line) how they were obtained.

Chapter 2

1. ADCs with unequal step sizes are used in some specialized applications. They are not used in standard data acquisition/analysis systems.

2. In fact, if the input voltage is exactly 5.000 V then levels 2,047 and 2,048 are equidistant from it to within less than a μV and either code is possible. In practice, unless the ADC is calibrated very carefully the code returned for an input of exactly 5.000 V will be somewhere close to 2,047 or 2,048 but is not guaranteed to be either of these.

3. An electroencephalogram (EEG) is a recording of the very small voltages produced by the brain. The occipital region of the brain (at the back of the head) processes visual information.

4. The exact number of bytes depends on the accuracy required and the type of computer used.

5. CMOS (pronounced 'see-moss') stands for complementary metal oxide semiconductor, which describes how the integrated circuits are made.

6. TTL stands for transistor–transistor logic, which also describes how the integrated circuits are made.
7. There are several varieties of this transistor. The 2N2222 has a metal case. The PN2222 has a plastic package that is more convenient in dense circuit boards. The PN2222A/2N2222A have a slightly different voltage rating. Any of the varieties will work in this circuit.

Chapter 3

1. A 'signal' is something that carries information. In the context of this book a signal is a voltage that changes with time and represents useful information, for example, blood pressure or temperature.
2. Poles and zeros are terms used in the theoretical analysis of filters. The operation of the filter can be described by its transfer function, which relates the input and output of the filter in complex (magnitude and phase) form. The transfer function can be expressed as an algebraic fraction whose numerator and denominator can be factorized. Factors of the numerator are zeros and factors of the denominator are poles. Pole–zero analysis is used extensively both to describe filter performance and design new filters.
3. As the number of poles increases so does the precision of the components needed to make the filter and guarantee that its performance will be within specification. At some point, standard resistors and capacitors are no longer adequate and high precision components have to be used. At this point the price increases rapidly because precision components can easily cost 100 times more than standard ones.
4. The electrocardiogram (ECG) is a recording of the electrical signals produced by the heart.
5. The electromyogram (EMG) is a recording of the electrical signals produced by muscles when they contract.

Chapter 4

1. Strictly speaking, strain is not quite the change in length divided by the original length. If you take a bar of something and stretch it by pulling on the ends, the bar gets longer. It is more correct to calculate the strain in the bar as the change in length produced by a small increase in the force applied to the ends divided by the actual (stretched) length of the bar, not the original length. Strain calculated using the actual length is called true strain, natural strain, or logarithmic strain, whereas strain calculated using the original length of the bar is called apparent strain. For materials such as metals and silicon, where practical strains are very small, the difference between natural strain and apparent strain is negligible. The difference becomes important when dealing with very stretchy materials such as rubber and biological materials.

Chapter 5

1. This is only true for noise which is random and uncorrelated with the signal, that is, there is no relationship between the signal and the noise. For the sort of noise found in biological signals (ECG, EMG, and amplifier noise) this is a reasonable assumption.

2. With an analytical method the equation of the best-fit curve can be stated in algebraic form. Given any particular data set, the best-fit curve is then calculated simply by plugging the data values into the equation.

3. A complex number has two parts, a real part and an imaginary part. The real part is just a normal number and can be positive or negative. The imaginary part is a real number times the square root of -1. $\sqrt{-1}$ is denoted by either i in mathematical work or j in engineering studies (since i is conventionally used to denote electrical current). Examples of complex numbers are: $5 + 3j$, $-1 + 2j$, $0.07 - 1.78j$. The two parts (real and imaginary) of a complex number are different types of numbers and cannot be combined. You can think of them as apples and oranges: you can add oranges to oranges and apples to apples, but you cannot add apples and oranges. In column 4 of Table 5.3 the fact that each imaginary number is multiplied by j is understood: the complex number in the first row, columns 3 and 4, would be written formally as $21.970728 + 0j$. Complex numbers may seem abstract but they have some important properties that make them very useful in signal processing. The theory of complex numbers is beyond the scope of this book but it is covered in standard mathematical texts.

4. A DC ('direct current') signal is one that does not change with time, that is, it is simply a constant voltage.

5. A sine wave is defined by the equation $V = a\sin(2\pi ft)$, where V is the voltage at time t, f is the frequency and a is the amplitude. Thus V varies from $+a$ to $-a$ volts. It is often easiest to measure the peak-to-peak height of a sine wave, and the amplitude is then half the peak-to-peak height.

6. From now on we will only plot the first half of the frequency domain, which is the convention when dealing with transforms of real signals.

7. There is also a hanning window which is very similar to the Hamming window. To add to the confusion the hanning window is also called the Hanning, Hann or von Hann window.

8. Meyer waves are slow oscillations in blood pressure and heart rate caused by blood pressure regulation mechanisms.

9. This was calculated as power spectrum (dB) $= 10\log_{10}$(power spectrum). Thus the dB scale is relative to $0\,\text{dB} = 1\,\text{ms}^2\,\text{Hz}^{-1}$.

10. TSAS: available by anonymous FTP from *psas.p.u-tokyo.ac.jp*. The README file gives more details about downloading the main files. TSAS was written by Dr. Yoshiharu Yamamoto (Laboratory for Exercise Physiology and Biomechanics, The University of Tokyo, 7-3-1 Hongo, Bunkyo-ku, Tokyo 113, Japan; *yamamoto@educhan.p.u-tokyo.ac.jp*)

11. 'Screened' or 'shielded' cables have a layer of braided copper or metal foil wrapped around the central conductor. If the shield is grounded it

blocks almost all capacitative coupling. Magnetic fields penetrate copper well and thus screening does not help prevent magnetically coupled interference. It is possible to shield against magnetic fields using high permeability alloys ('mu-metal') but these are not totally effective and are only used to shield objects. They are not used in cables.

12. There are several ways to represent a colour image. The red, green, blue (RGB) system relates directly to the way that an image is formed on a computer monitor or by a projector. Each colour is made up by adding together red, blue, and green light. The cyan, magenta, yellow (CMY) system is used in printing where colour is formed by removing different wavelengths from white light. This system does not give a good black on printed paper and the latter is added to produce the four-colour CMYK system. The hue, saturation, intensity (HSI) system is closest to the way that colours are appreciated by the human visual system.

13. NIH Image and related information are available by anonymous FTP from *zippy.nimh.nih.gov*. There is an active discussion group on its use. To join the list send the email message: subscribe nih-image <your real name> to *listproc@soils.umn.edu*

14. In the NTSC system each line is 63.5 µs long but only 52 µs is used for picture information. A typical video digitizer divides each line into 640 samples so the sampling rate is 12.3 MHz.

15. Most video digitizers are eight-bit.

Chapter 6

1. Imagine a tank of coloured water that is full to the brim. A tap adds fresh water to the tank at a constant rate and an equal amount of coloured water overflows out of the tank. If we assume that the water from the tap is mixed rapidly with the water in the tank then the dye concentration in the tank falls exponentially with time.

2. EEG signals are a few µV in amplitude and it is difficult to avoid some power line interference when recording them.

3. The spectrum of a practical noise signal has to decrease at some point, otherwise the signal would have infinite energy.

4. Matlab was used to calculate the Fourier transform. The algorithm in Matlab looks at the number of data points and uses an FFT of appropriate radix if it can. If not, it defaults to the DFT but in either case it uses the whole data set. The practical advantage of using this algorithm is that it guarantees to transform the whole data set in the most efficient manner without the user having to worry about the details.

5. A lot of the terms used in electrophysiology come from electronics. Until recently all the signal processing for electrophysiology was performed by analogue circuits and a lot is still done this way. The circuits are described by their function: amplifier, integrator, rectifier, and so forth.

References

Barahona, M. and Poon, C.-S. (1996) Detection of nonlinear dynamics in short, noisy time series. *Nature* **381**, 215–17.

Bracewell, R. N. (1984) The fast Hartley transform. *Proc. IEEE* **72**, 1010–18.

Cooley, J. W. and Tukey, J. W. (1965) An algorithm for the machine calculation of complex Fourier series. *Mathematics of Computation* **19**(90), The American Mathematical Society, Providence, RI, USA.

Hartley, R. V. L. (1942) A more symmetrical Fourier analysis applied to transmission problems. *Proc. IRE* **30**, 144–50.

Horowitz, P. and Hill, W. (1989) *The art of electronics*. Cambridge: Cambridge University Press, second edition.

Kay, S. M. and Marple, S. L. (1981) Spectrum analysis – A modern perspective. *Proc. IEEE* **69**, 1380–419.

Lynn, P. A. (1992) *Digital signals, processors and noise*. Basingstoke: The Macmillan Press Ltd., first edition.

Nyquist, H. (1928) Certain Topics in Telegraph Transmission Theory. *Trans. A.I.E.E.*, April, 617–44.

Pan, J. and Tompkins, W. J. (1985) A real-time QRS detection algorithm. *IEEE Transactions on Biomedical Engineering* **BME32**(3), 230–6.

Press, W. H., Teukolsky, S. A., Vettering, W. T. and Flannery, B. P. (1992) *Numerical recipes in C: the art of scientific computing*. Cambridge: Cambridge University Press, second edition.

Rabiner, L. R. and Gold, B. (1975) *Theory and application of digital signal processing*. Englewood Cliffs: Prentice-Hall Inc.

Russ, J. C. (1994) *The Image Processing Handbook*. Boca Raton: CRC Press, second edition.

TenVoorde, B. J., Faes, T. J. C. and Rompelman, O. (1994) Spectra of data sampled at frequency-modulated rates in application to cardiovascular signals: Part 2. Evaluation of Fourier transform algorithms. *Med. Biol. Eng. Comput.* **32**, 71–6.

Yamamoto, Y. and Hughson, R. L. (1991) Coarse graining spectral analysis: a new method for studying heart rate variability. *J. Applied Physiology* **71**, 1136–42.

Index